McGraw-Hill Networking Telecommunications

Build Your Own
Trulove — *Build Your Own Wireless LAN with Projects*

WITHDRAWN

Crash Course
Louis — *Broadband Crash Course*
Vacca — *i-Mode Crash Course*
Louis — *M-Commerce Crash Course*
Shepard — *Telecom Convergence 2/e*
Shepard — *Telecom Crash Course*
Louis — *Telecom Management Crash Course*
Bedell — *Wireless Crash Course*
Kikta/Fisher/Courtney — *Wireless Internet Crash Course*

Demystified
Harte/Levine/Kikta — *3G Wireless Demystified*
LaRocca — *802.11 Demystified*
Muller — *Bluetooth Demystified*
Evans — *CEBus Demystified*
Bayer — *Computer Telephony Demystified*
Hershey — *Cryptography Demystified*
Taylor — *DVD Demystified*
Hoffman — *GPRS Demystified*
Symes — *MPEG-4 Demystified*
Camarillo — *SIP Demystified*
Shepard — *SONET / SDH Demystified*
Topic — *Streaming Media Demystified*
Symes — *Video Compression Demystified*
Shepard — *Videoconferencing Demystified*
Bhola — *Wireless LANs Demystified*

Developer Guides
Vacca — *i-Mode Crash Course*
Guthery — *Mobile Application Development with SMS*
Richard — *Service and Device Discovery: Protocols and Programming*

Professional Telecom
Smith/Collins — *3G Wireless Networks*
Bates — *Broadband Telecom Handbook 2/e*
Collins — *Carrier Class Voice Over IP*
Harte — *Delivering xDSL*
Held — *Deploying Optical Networking Components*
Minoli — *Ethernet-Based Metro Area Networks*
Nagar — *Telecom Service Rollouts*
Benner — *Fibre Channel for SANs*
Bates — *GPRS*
Bates — *Wireless Broadband Handbook*

Sulkin	*Implementing the IP-PBX*
Lee	*Lee's Essentials of Wireless*
Bates	*Optical Switching and Networking Handbook*
Wetteroth	*OSI Reference Model for Telecommunications*
Russell	*Signaling System #7 4/e*
Minoli	*SONET-Based Metro Area Networks*
Louis	*Telecommunications Internetworking*
Russell	*Telecommunications Protocols* 2/e
Minoli	*Voice over MPLS*
Karim/Sarraf	*W-CDMA and cdma2000 for 3G Mobile Networks*
Bates	*Wireless Broadband Handbook*
Faigen	*Wireless Data for the Enterprise*

Reference

Muller	*Desktop Encyclopedia of Telecommunications* 3/e
Botto	*Encyclopedia of Wireless Telecommunications*
Clayton	*McGraw-Hill Illustrated Telecom Dictionary* 3/e
Pecar	*Telecommunications Factbook* 2/e
Russell	*Telecommunications Pocket Reference*
Radcom	*Telecom Protocol Finder*
Kobb	*Wireless Spectrum Finder*
Smith	*Wireless Telecom FAQs*

Security

| Nichols | *Wireless Security* |

Telecom Engineering

Smith	*Cellular System Design and Optimization*
Rohde/Whitaker	*Communications Receivers* 3/e
Sayre	*Complete Wireless Design*
OSA	*Fiber Optics Handbook*
Lee	*Mobile Cellular Telecommunications* 2/e
Bates	*Optimizing Voice in ATM/IP Mobile Networks*
Roddy	*Satellite Communications* 3/e
Simon	*Spread Spectrum Communications Handbook*
Snyder	*Wireless Telecommunications Networking with ANSI-41* 2/e

BICSI

Network Design Basics for Cabling Professionals
Networking Technologies for Cabling Professionals
Residential Network Cabling
Telecommunications Cabling Installation

IP Telephony Demystified

Ken Camp

McGraw-Hill
New York   Chicago   San Francisco   Lisbon
London   Madrid   Mexico City   Milan   New Delhi
San Juan   Seoul   Singapore   Sydney   Toronto

The McGraw-Hill Companies

Library of Congress Cataloging-in-Publication Data

Camp, Ken.
    IP telephony demystified / Ken Camp.
        p. cm.
        ISBN 0-07-140670-0 (alk. paper)
        1. Internet telephony.   I. Title.
    TK5105.8865 .C34 2002
    621.382'12—dc21

                            2002032669

1 2 3 4 5 6 7 8 9 0   DOC/DOC   0 9 8 7 6 5 4 3 2

ISBN 0-07-140670-0

*The sponsoring editor for this book was Stephen S. Chapman and the production
supervisor was Sherri Souffrance. It was set in New Century Schoolbook by
Patricia Wallenburg.*

Printed and bound by RR Donnelley.

McGraw-Hill books are available at special quantity discounts to use as premiums
and sales promotions, or for use in corporate training programs. For more information, please write to the Director of Special Sales, Professional Publishing,
McGraw-Hill, Two Penn Plaza, New York, NY 10121-2298. Or contact your local
bookstore.

This book is printed on recycled, acid-free paper containing a minimum of
50 percent recycled, de-inked fiber.

For Pat,

Thank you.

CONTENTS

Preface   xi
Acknowledgments   xv

Chapter 1   History and Overview of Telecommunications   1

Public Switched Telephone Network   3
Structure of the Telephone Network   5
    The Local Loop   6
    The Central Office Switch   7
    Components of the PSTN   7
Analog versus Digital Signals   9
    Analog Signals   9
    The Sine Wave   10
    Digital Signals   11
    Transmitting Analog and Digital Signals   11
Bandwidth versus Passband   12
    Bandwidth in the PSTN   14
    Harry Nyquist and Signaling Rate   14
    Bit Rate versus Baud   15
The Impact of Noise on Signals   15
    Claude Shannon's Theorem   16
Switching   17
    Circuit Switching   17
    Packet (Store-and-Forward) Switching   19
Connectionless versus Connection-oriented Networks   20

Chapter 2   The Internet and IP Networking   23

OSI Network Reference Model   24
    The Physical Layer   25
    The Data Link Layer   26
    The Network Layer   26
    The Transport Layer   27
    The Session and Presentation Layers   28
    The Application Layer   28
The Importance of TCP/IP and the Internet Protocol Suite   28
    The TCP/IP Suite   29
    Internet Protocol (IP)   30

IP Addresses   34
    Acquiring an IP Address   37
    Dynamic Addressing   38
    Private Addresses and the Public Network   40
Reliability versus Quick Delivery   41
    Addressing the Layers   42
    Transmission Control Protocol (TCP)   45
    User Datagram Protocol (UDP)   49

Chapter 3   Fundamentals of Packetized Voice   53
A Word About the Standards   54
A Telephone Call Simplified   56
Analog to Digital Conversion   57
Pulse Amplitude Modulation (PAM)   58
The Complete Digitizing Process   61
IP Telephony versus Traditional Telephony   62
Voice Quality in the Network   64
When Does Packetized Voice Become Telephony?   67

Chapter 4   IP Telephony Protocols   71
H.323 Standards for Multimedia Over Packet Networks   74
Call Setup Using H.323   77
Session Initiated Protocol (SIP)   80
Session Description Protocol (SDP)   82
Call Setup Using SIP   83
Comparison of H.323 and SIP   85
Megaco and H.248   86
    Megaco Terms and Definitions   89
    The Gateway Commands   89
    Call Setup Using Megaco/H.248   92
Real-time versus Nonreal-time Traffic   94
Which Protocol is Needed? Which is Best?   94

**Chapter 5   The Lure of IP Telephony:
                Can It Work for Me?   97**
Voice versus Data   98
The Cost of Doing Business   99
Telephony in the PSTN versus IP Telephony   101
When Does It Become IP Telephony?   104
Considerations for Business Evaluations   106
    Cost Reduction   106
    Increased Revenue   107
    Increased Efficiency   107
    Customer Satisfaction Improvement   108

Access to New Market Share/Market Segment   108
Provide New Service   108
The Silver Bullet   109
IP Telephony in the Enterprise   111
IP Telephony in the Mid-Sized Business   113
IP Telephony in the Small Business   115
Debunking Myths   116
Preparing to Implement an IP Telephony
Solution—Planning for Success   117

**Chapter 6   Performance, Quality of Service,
and Traffic Engineering   121**

Quality of Service Parameters   124
Finding the Network Performance Envelope   127
Acceptable IP Telephony   131
Quality of Service within IP   131
Three QoS Philosophies for IP   133
Integrated Services (IntServ)   134
Resource Reservation Protocol   135
Differentiated Service (DiffServ)   138
Multiprotocol Label Switching (MPLS)   139
Where's the "Quality"?   142
Gigabandwidth   143

Chapter 7   Gateways between Networks   145

Why Gateways Are Necessary   146
The Service Provider Basic Model   147
The ILEC Gateway   148
The ITSP Gateway   149
The Enterprise Model   152
Types of Gateways   155
Gateway or Softswitch?   156

Chapter 8   Premise-based Telephony   161

Digital Subscriber Line   165
Cable Modems   169
Satellite Internet   171
Fixed Wireless Internet Technologies   173
WiFi   174
If We Build a High-Speed Network, Will They Come? 176
Work-at-Home Staff—Telecommuting   177
Replacing the PBX in the Office   179
Computer Telephony Integration (CTI)   184
Why Not Implement All Three?   186

Chapter 9   Network-based IP Telephony Services   187

Traditional Centrex Service   189
    Centrex Features   190
Centrex Evolves to IP Centrex   193
    IP Centrex in the Class 5 Switch Architecture   194
    IP Centrex in the Softswitch Architecture   195
    Customer Premises Equipment for IP Centrex   196
    IP Centrex Benefits   198
Does IP Telephony Include the Internet?   201
Privacy: Fact or Fiction?   201
Voice Over DSL/Frame/ATM—Variations on a Theme   202
    Voice over DSL (VoDSL)   203
Services Perspective   206

Chapter 10   The Future of IP Telephony   207

Beefing Up IP Telephony   210
Industry Changes Affecting IP Telephony   212
Disruptive Technology Takes Center Stage   213
Internet Call Centers   216
Fax Over IP   220
Voice Mail over IP   222
IP Telephony and Denial of Service (DoS) Attacks   224
The Future for Equipment Vendors   225
The Future for Telecommunications Service Providers   227
The Future for Business Users   229
The Evolution Continues   230

Appendix A   Glossary   233

Appendix B   Resource List   239

Solution Vendors   239
Standards Organizations   241
Newsletters and Information Sites   241
General Interest URLs   242

Index   243

PREFACE

Every business in the world uses telecommunications, but this technology has introduced an incredible number of problems as well. Business telephony ranges from the small office with one or two phone lines, to large campus environments with multiple PBXs linked together, and every possible permutation in between.

In today's market, essentially everyone uses the Internet as a business tool. We've heard the term convergence bandied about in many ways, but in essence, we speculate about the feasibility and reality of integrating all services, voice, data and video, onto a single network infrastructure. This level of convergence, while possible, is for many companies, not practical today.

What may be practical, realistic, and even profitable, is integrating voice and data onto a single IP infrastructure. Initially, the perceived benefit of this integration was reduced toll and long distance charges. Over time, as the wireless market has grown, we've seen long distance costs reduced significantly over the past few years.

Today the sensible use for IP telephony may be integration of services. Integrating voice and data onto a single infrastructure can provide cost savings by reducing the overhead, in both equipment and talent, required to provision and maintain these two mission-critical services. Beyond that, integration of voice and data leads to new services and customer service offerings that are only beginning to be explored. The voice-enabled call center is merely the tip of the "integration iceberg." We must integrate in order to innovate, and IP telephony presents an area in technology where we have all the tools within our grasp to succeed.

This book explores both the business concepts and the technologies at levels that will help both business managers and technologists make intelligent and informed decisions about why IP telephony might provide value to their companies. All too often technology books explore the bit level detail of a technology without explaining why the reader should know or care about the core technology. We'll dive into the technology in several areas and uncover the underlying importance of both the technology and how it can be implemented, but we'll also explore some of the reasons why it might make sense and look at some well-informed approaches to implementing an integrated solution.

The telecommunications providers of the world have often been slow to migrate to new technologies, and IP telephony is no different. When you hold the lion's share of the market, there isn't always strong motivation to innovate. Both IP telephony and xDSL are viewed by many people as disruptive technologies, but DSL isn't truly disruptive to the telecommunications industry.

The DSL market has seen many failures as the CLECs tried to compete and discovered both technical and business impediments to success. The ILECs individually chose to either explore or ignore the technology in many instances. No matter which approach chosen, they have collectively failed to embrace DSL as a profitable and desirable market segment. Market penetration statistics bear this out. For many reasons, not the least of which is poor legislation, DSL has not delivered the bandwidth necessary for many integrated consumer services.

IP telephony has a fundamental difference from DSL. For DSL to succeed, the telecommunications service providers had to do something enormously different than they had previously been doing. The Telecommunications Act of 1996 and the issues surrounding the unbundled local loop required the incumbent telco providers to make fundamental changes in their mindset. These changes weren't likely, and never came. IP telephony on the other hand, can succeed in the market even if the telcos continue to operate the way they always have and change nothing. In fact, the more the telcos drag their feet, the greater the likelihood that they will become the victims of disruption. Because IP telephony is a service riding over existing transport technologies, readily available, new companies can enter the market without concerted efforts by the telcos to help make the technology a success. In this regard, IP telephony is a true disruptive technology in that, if ignored by the incumbents, it can grow to overwhelm the existing technology base.

During the writing of this manuscript, the industry changed yet again. Bankruptcies and legal problems have hit more of the major carriers. The industry has been attacked by the media and abandoned by investors. The shift in the financial climate serves as motivation to the remaining players to maintain the status quo. Make no mistake, the telecommunications industry, particularly the incumbent traditional providers, are treading through a minefield as they balance the need for new services against sound business tactics. The motivation to continue doing what they have always done will leave more casualties behind as technology moves forward. Don't let your business become a casualty in an environment where the only constant is change.

Ken Camp
ken@ipadventures.com

ACKNOWLEDGMENTS

For technical and professional assistance in developing the content of this manuscript, I wish to thank the following people. Without their vast expertise and understanding of how technologies fit into life today, the book would not be as complete as it is. While there have been too many to count over the years, I'm especially grateful to Jerry Johnson and Alan Kamman at the Vermont Telecom Advancement Center, who compel me to always consider the "little guy"; Mitch Moore of Globix Corp. a keen mind with insights in many areas; Todd Pritsky, Ed Seager, Mark Steinberg of Hill Associates, along with many former colleagues there, including Karen Andresen and Melody Vance who always offer sound technical thoughts and kind words of encouragement.

I owe a special thanks to Joe Passafiume of MultiSystems Interconnect, Inc. We work together on a good many projects, but Joe graciously allowed me to use and discuss some of his visual representations in the Quality of Service chapter. Joe's understanding of network performance issues is unsurpassed, and I count him as both a professional colleague and personal friend.

Acknowledging the Internet for the fabulous resource it has become may be ludicrous, and for many of us, resources like the Internet Engineering Task Force are just a click away. It's important to acknowledge some of the companies that make good information available, including 3Com, Alcatel, Cisco Systems, Intel, Lucent Technologies, Microsoft, Nortel Networks, Shoreline Communications, and Voyant Technologies.

For their help with the manuscript, I'm indebted to my wife Pat for both moral and editorial support, and Patty Wallenburg for her editing support. I'm grateful to Steve Shepard, a colleague who trod this path before, and encouraged me to make the leap, and to Steve Chapman at McGraw-Hill who saw the potential in the outline that became this book.

I'm deeply grateful to my mother and to my wife. Their love, support, encouragement, and understanding has helped me forge ahead, even in bleak times.

History and Overview of Telecommunications

To understand and appreciate some of the business and technical issues driving the popularity of carrying voice over the Internet protocol (IP), we need to begin with some basic framework of how the telephone system works and how it came to be. This chapter provides a basis of telecommunications technology and history to give us better insight into why some issues exist that we now perceive as problems, and why IP telephony might be a solution that makes good sense in today's network environment.

It's been argued for years that humans are very social creatures. Early nomadic tribes met to trade, and cultures were shared by that contact. As society evolved, roads connected larger communities, then smaller towns and villages. Many technologies have evolved over time to bridge the gap between humans and change how we interact. The car displaced the horse and drove the creation of a national road system. The airplane has become a typical mode of travel for reaching destinations far away.

Communications has evolved as well (Figure 1.1). Early man used pictographs on cave walls to express ideas. Today, we pick up the phone and call someone on the other side of the world. The delivery of our information or idea is nearly instantaneous. Our culture has shaped us

Figure 1.1
Evolution of human communication.

to expect things to happen very quickly, and we have become impatient about delays in delivering information in any form.

To appreciate the technologies of today, we must consider some of the evolution that brought us here. The evolution of communications has been filled with wondrous advances in our ability to share information and ideas.

The written word evolved, as did spoken language. In the mid-1400s, Johannes Guttenberg's invention of the printing press using movable metal type opened the door for mass communications like magazines and newspapers, but person-to-person communications was still conducted face-to-face or via handwritten messages.

The telegraph was developed for communication over electrical wires, yet was constrained to brief messages. Longer messages still had to be conveyed in a letter, whether typed or handwritten, then delivered via some carrier method.

Public Switched Telephone Network

On March 10, 1876, Alexander Graham Bell made the first telephone call, and we were given a glimpse of a technology that would spread around the world and change how we communicate. The public switched telephone network (PSTN) has been growing ever since. In 1915, AT&T opened the first transcontinental phone lines for business, thus enabling telephone calls across the country.

In the years that followed, the telephone changed how we communicate, how we share information, even how we talk. According to stories, Alexander Bell preferred to answer the telephone *Ahoy*, whereas it was Thomas Edison who arguably gave us *Hello* as a greeting.

Some of the most significant events in the evolution of the telephone network weren't directly related to technology. When consumers were offered flat-rate pricing for local calling, usage of the telephone rose dramatically. In the mid-1960s, AT&T offered direct-dial long distance, and for the first time we could call friends or family across the country without the intervention of a long-distance operator. Again, this precipitated a dramatic rise in telephone calling patterns.

To transmit a telephone conversation, we must send sound over long distances. In a telephone call, this sound is voice, but it could just as well be music or some other sound. The design of the PSTN over the past century has focused on carrying voice traffic in the most efficient way possible using the engineering techniques available.

Humans produce sound with their vocal cords. The wind creates sounds. An object being dropped makes a sound when it hits the floor. The sound we hear is a set of variations in air pressure set up by vibration. Air is the medium that carries the sound. Much like ripples on a lake, sound waves travel through the air.

Just like the ripples in the lake, the farther from the source the sound waves travel, the weaker they become, because energy is being expended to carry the sound over distance. As energy is lost, volume decreases, because the vibrations are weaker. This loss of energy equates to a loss of signal strength as the sound is transmitted. In practical terms, most sound can be transmitted about 3,000 feet.

The invention of the telephone set used in that first call signaled the beginning of a new era. It provided a way to convert the physical energy of sound waves or vibrations into electrical energy that could be transmitted over copper wires. Loss of energy still occurs, but electrical energy can use *repeaters* or *amplifiers* to extend the distance a signal can be transmitted.

The telephone handset contained a microphone in the mouthpiece that was powered by the telephone network. This microphone was filled with carbon granules and connected in series using electrical potential. As the human voice sets up vibrations, the granules compress and expand in response to the changes in air pressure. As Ohm's law dictates, voltage, current, and resistance are all related, permitting the electrical circuit to produce a signal with a variable current that models the sound wave.

This electrical signal can be transmitted over wires or a circuit to another telephone. At the receiving end, the electrical impulses are delivered to a speaker through the earpiece. The earpiece contains an electromagnet and metal disc. The variations in the electrical current and state changes in the electromagnet set up vibrations in the disk that are heard as an audible signal by the human ear. This vibration generates sound waves that the ear and brain hear as the voice of the caller.

This component in the telephone set is referred to as a *transducer*. It's the mechanism for converting sound waves to electrical impulses and then back into sound waves. As long as the integrity of the electrical signal is maintained, it can be transmitted over very great distances.

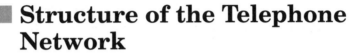

Structure of the Telephone Network

One way to create a telephone network is to connect a set of wires from every subscriber, to every other subscriber. This would provide a dedicated connection for every possible conversation, but the idea of physical connections from any user to any user creates a major engineering dilemma. In network parlance, we use the formula shown in Figure 1.2, where n equals the number of nodes or users on the network.

Figure 1.2
Using this formula 5,104,641,789,979,500 pairs of wire would be needed to connect every household in the United States. This only represents residential users, not business.

Formula to Calculate Number of Connections

$$\frac{n(n-1)}{2}$$

Calculations based on Latest Census

$$\frac{101,041,000(101,041,000-1)}{2}$$

To overcome this impossible task as the telephone network grew, the industry built a switched network, which allowed connections to be established for the duration of the phone call, then disconnected at the end of the conversation. Using this approach, users have a single pair of wires connecting their telephone to the network, sharing the equipment and resources inside the network, and only using them when actually making a telephone call. This approach provides reasonable service availability, based on documented calling patterns and appropriate network design by the phone companies. Since everyone doesn't need to talk at the same time, the telephone network doesn't have to support a full-time connection everywhere. This network lets anyone talk to anyone, but allows cost-effective sharing of the central office (CO) equipment. As end-users, we get reasonable service at a reasonable cost to the telephone company.

The components of this switched telephone network are shown in Figure 1.3. Each customer is connected to the CO by a pair of copper wires, referred to as the *local loop*. It isn't always a pair of wires connected clear back to the central office (CO), although that was an early common arrangement.

The equipment in the CO allows any subscriber's connection to be connected, or switched, to any other connection. Each CO serves a *radius*, or geographic serving area, that allows transmission of voice calls with reasonable quality. Central offices are connected using *trunks*, or lines capable of handling more than one phone call. If a trunk has 24 channels, it can handle 24 connections between the COs. Once a connection is dropped, another user can reuse the trunk to place a call, so the lines are shared sequentially, based on need.

Figure 1.3
Components of a switched telephone network.

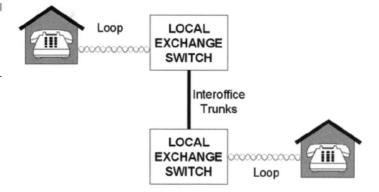

The Local Loop

The local loop (Figure 1.4) is the connection between the customer or subscriber and the CO switch. Because this is the customer's access to the network, it is commonly dedicated to a single subscriber and available to make a call whenever the subscriber wishes. It consists of a pair of copper wires, sometimes referred to as a *two-wire circuit* or *twisted pair*. It's important to remember that this technology was developed and matured before computers came into everyday use; because the network has been optimized for many years to transport voice calls, the focus on voice service is one of the factors that has caused difficulty in deploying services like digital subscriber line (DSL).

Figure 1.4
The local loop.

Dedicated connection between subscriber and network
•Two wires (twisted pair) for economics
•Signals travel in both directions over wires

The Central Office Switch

The CO switch sets up the temporary connections or circuits between users making a telephone call. To manage this, local loops and trunks all terminate at the CO switch. The switch also has service circuits used for the supervision and management of the network itself. These circuits monitor call setup and disconnects, send announcements or signals, and gather information. Older switches used electromechanical relays. More modern switching equipment is all digital in nature, using solid-state electronics—making the CO switch, in essence, a large computer optimized for handling telephone calls. These switches are placed in COs or wire centers that also contain battery and power equipment, test boards, frame terminations for all the equipment and cabling, and a cable vault where all the cabling enters the building.

Components of the PSTN

Figure 1.5 shows how the different major components in the PSTN connect and interact. The sequence of events that takes place when a subscriber makes a call is as follows:

1. A call originates at a subscriber's premise using the twisted pair local loop to connect to the CO. Several variations of telephone service can be provided over twisted pair, including plain old telephone service or POTS, digital services like the integrated services digital network (ISDN), and new high-speed data services like DSL technologies. Many other services, such as traditional T-1 circuits, require two pairs of wire.

2. The end offices provide connections to local users within their serving area, and the local switch handles all connections that don't leave the area. When a user makes a call to a subscriber in another area, the switch connects to a trunk to extend the call outside the local serving area.

3. Trunks are *four-wire circuits*, using two pairs of wires. This two-wire and four-wire terminology is a carryover from the early days of the telephone network, when all connections were made using pairs of copper wires. The need for two pairs of wires was for traffic in both directions at the same time, or *full-duplex* communications. Today, these connections are often made using fiber connections. Trunks are usually deployed in groups of circuits, so that a trunk group might consist of 100 or more connections between two CO switches.

4. A *tandem switch* is used to connect offices together. As the PSTN grew, it became impractical for every CO to have connections to every other CO, so the tandem switch became a concentration point for trunks.

5. A *point of presence* or POP is a tandem switch that a long distance carrier or *competitive local exchange carrier* (CLEC) owns and uses to connect into a serving area.

6. *Signaling System 7* (SS7) is a key component of the network. SS7 is a network that overlays the PSTN and connects databases and switches in the network, along with other network elements. This secondary network provides the intelligence used for call setup and enhanced services, such as 800 services and credit card billing. SS7 is the core of the advanced intelligent network (AIN).

Figure 1.5
Components of the public switched telephone network.

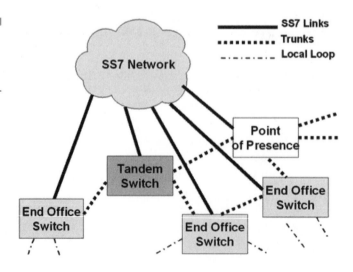

Analog versus Digital Signals

A signal represents some form of data on some form of physical medium. Signals that are continuously variable are commonly thought of as being analog, whereas discrete signals are considered digital. For example, water pressure from a hose would be an analog signal, whereas flipping a light switch off and on would represent a digital signal. Generally, digital signals have predefined levels that represent specific signals.

Analog signals are represented by a waveform or *sine wave*, as shown in Figure 1.6. This waveform allows representation of the all the possible variations.

A digital signal can only represent predefined values. In a binary system, there are two levels, for example on and off, but many digital systems have a set of predefined values. Some would argue that the English alphabet consists of 26 digital values represented by the letters. So, a digital system doesn't necessarily have to be a binary system.

Figure 1.6
Analog and digital signals. Analog signals are represented by a sine wave.

Analog - Variable signals across an infinite set of values

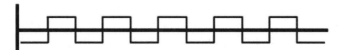

Digital – Discrete signals from a predefined set of values

Analog Signals

Figure 1.7 shows two examples of analog signals: one generated by striking a key on a piano, causing the hammer to strike the string, and the other by the golf ball bouncing down the fairway. Because we use the sine wave to describe analog signals, we need to understand some properties associated with sine waves.

Figure 1.7
Analog signals.

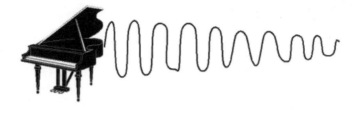

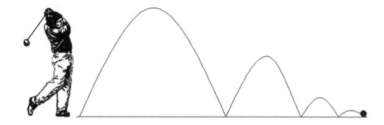

The Sine Wave

A sine wave (Figure 1.8) is a mathematical construct. The string of a guitar is a good example of how a wave is generated. When the string is plucked, it moves from its taut stationary position and this motion is measured over time. In telephone networks, we must consider three properties of the sine wave: amplitude, frequency, and phase.

Amplitude is the height of the wave in relation to a horizontal axis. In Figure 1.7, we could create two waves of differing amplitude by using a piano key. Striking the key softly would create a wave with lower amplitude than striking the key hard. Amplitude is related to the strength of the vibration or *volume*, when referring to sound.

Frequency is a correlation to the number of cycles a wave repeats within a time interval. A typical measurement of frequency uses *cycles per second* or *Hertz* (Hz). The human ear hears different frequencies at different pitches, so different keys on the piano, which strike different strings inside, set off different frequency vibrations, or notes, to our ear.

Phase is the relationship in time to when the vibration starts. If two piano keys played the same note, and one was struck a half-cycle after the other, the sine waves would be 180 degrees, or a half cycle *out of phase* with one another.

Figure 1.8
Sine wave.

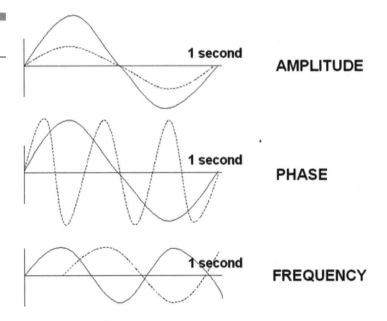

Digital Signals

Like an analog signal, a digital signal can be described as a function that varies over time, but can only contain a discrete set of values. Figure 1.9 shows a square wave varying between the two distinct and discrete values possible in this example.

The square wave is very similar to a sine wave, two pulses (up and down) to represent a cycle. Using this approach, a square wave that pulses up or down in value 8,000 times per second would have a frequency of 4,000 Hz.

The square wave is not a sine wave, but it has enough similarities to suggest a relationship between the two. The relationship was completely described by mathematician Jeanne-Baptiste Fourier.

Transmitting Analog and Digital Signals

Any signal loses strength over the distance it is transmitted. We call this *attenuation* or *loss*. Analog and digital transmission methods are both subject to attenuation regardless of the type of data being transmitted, and each method deals with loss in its own way.

Figure 1.9
Square wave versus
sine wave visual.

Is There a Relationship?
•Both are a funtion of time
•Analog has continuous range of values
•Digital has a discrete set of values

Analog signals overcome loss by amplifying the signal along the way. *Amplifying* means simply increasing the amplitude. When a weak signal enters an amplifier, it is boosted in strength and transmitted back out. In analog transmission, no effort is made to distinguish the actual signal from any noise that might be present. Both the signal and the noise are amplified and transmitted.

Digital transmission methods assume that the signal is carrying digital data. A device called a *repeater*, which reads the weak signal and regenerates the original signal from that information, can handle loss in signal strength. In Figure 1.10, the square wave at the top represents a T-1 digital signal leaving the transmitter. As we move farther and farther from the transmitter, notice how the signal degrades due to attenuation. A repeater takes the weakened signal, shown at the bottom of the figure and regenerates the signal as it was originally sent.

Because this regeneration process doesn't recreate or amplify the noise, noise is reduced on the circuit and creates a perception among many that digital transmission is noise free. There are, in fact, other types of noise that can cause problems on a digital transmission, but we don't need to explore them here.

Bandwidth versus Passband

Any channel used to transmit voice carries a range of frequencies. This range of frequencies is referred to as the *passband* of the channel. Normally, telecommunications services define passband as those frequen-

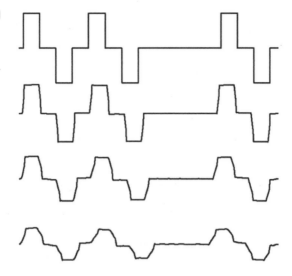

Figure 1.10
The degradation of
a digital signal.

cies that pass through the channel with at least half their power. See Figure 1.11.

Bandwidth describes the width of the passband. A channel with a passband of 300 to 3,300 Hz has a bandwidth of 3,000 Hz. Bandwidth correlates directly to the amount of information a channel can carry. A channel with a passband of 300 to 3,300 Hz and a channel with a pass-

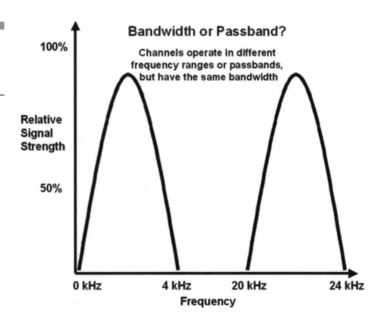

Figure 1.11
The relationship
between bandwidth
and passband.

band of 20,000 to 23,000 Hz have the same bandwidth (3,000 Hz), and effectively carry the same amount of information: bandwidth, not passband, determines the carrying capacity of a channel.

Bandwidth in the PSTN

The human voice can produce many frequencies, from about 50 Hz to more than 10,000 Hz. Earlier, we used the example of a passband of 300 Hz to 3,300 Hz for a reason: in the telephone network, voice channels are typically provided in 4 kHz increments. Our telephone connection to the CO uses from 0 Hz to 4,000 Hz as a voice channel, but only 300 Hz to 3,300 Hz are used for voice traffic. Testing determined long ago that frequencies in this range can be used to accurately transmit the human voice in conversation. This passband provides intelligibility of the signal and permits the receiver to recognize the person talking or transmitting. A noticeable difference exists between a voice heard over the telephone and a voice heard in the same room, because of the bandwidth available to transmit that voice, but this approach is used to conserve the bandwidth required in transmission systems within the telephone network.

A good example of how bandwidth affects the fidelity of a signal is found by comparing amplitude modulation (AM) and frequency modulation (FM) radio. AM radio uses about 5 kHz of bandwidth; FM radio transmits about 18 kHz. Thus, FM radio is able to transmit music with better sound quality. The bandwidth allows a richer signal with a higher fidelity.

Harry Nyquist and Signaling Rate

Working at Bell Labs in the 1920s, Harry Nyquist demonstrated that the maximum signaling rate of a channel, measured in *baud*, is directly related to the bandwidth of the channel. Nyquist showed that the maximum signaling rate is twice the bandwidth of the channel. In the telephone network, the passband of 300 Hz to 3,300 Hz gives us 3,000 Hz bandwidth. Using *Nyquist's theorem*, the maximum number of samples per second for this channel is 6,000 baud.

Because this implies that the signaling rate and *bit rate* are directly related to bandwidth, we now use the terms bandwidth and bit rate to mean the same thing. Today, there is much talk of higher-speed connectivity to the Internet, and we often hear people say they need more bandwidth than 56 kbps. What they mean is that they need a faster line speed. To achieve the faster line speeds, given our signaling rates and bandwidth, we need to delve deeper into the idea of bit rates and baud.

Bit Rate versus Baud

If every signal in a transmission conveys a single bit of information, then bit rate and signaling rate (baud) are equal. But what if each signal conveyed two bits of information to the receiver? Then the bit rate would be twice the signaling rate. As our technologies became more and more sophisticated, we learned to transmit more than one bit of information per signal. Thus, the bit rate equals the baud rate multiplied by the number of bits per signal.

To transmit more than one bit of information per signal, there must be enough discrete signaling states to allow every possible pattern that might be sent. To send two bits per signal, we might have to define 00, 01, 10, and 11 all as binary signal states. We can accomplish this by using different voltages to represent each value, as shown in Figure 1.12.

Figure 1.12
Bit rate versus
signaling rate (baud).

Suppose the signaling rate is 1,200 baud

Bit Rate	Bits per Signal	Number of Discrete Signals
1200	1	2
2,400	2	4
4,800	4	16

Bit Rate (bps) = signaling rate x bits per signal

The Impact of Noise on Signals

Noise is described as the electrical disturbance of a signal, usually random. It is the primary limiting factor on how many bits of information each signal can carry (see Figure 1.13). If there were no noise, the transmitted signal would arrive intact, but since there isn't a perfect channel, we must develop signal types that can be distinguished and interpreted even when noise is present. Although they must be distinguishable, they must also be different enough to be uniquely identifiable, thus limiting the number of signals we can have present.

To help clarify this, let's use colored ping-pong balls as signals. If we have a red ball represent a 1 and a pink ball represent a 0, if someone threw a ball at you, you could easily determine which signal was being sent. What if we used eight shades of red and each represented three bits of information (000,001,010, 011,100,101,110, and 111)? If someone throws a ball at you now, how easy is it to determine which shade of red it is? What happens if the signals come more quickly? Clearly, your ability to identify which signal is being transmitted is affected. What if we dim the lights in the room, or move the person throwing farther away?

The same holds true in the transmission of data. In our example, noise would be comparable to dimming the lights or moving the thrower farther away.

Figure 1.13
The effect of noise on a signal.

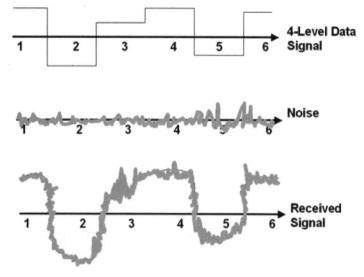

Claude Shannon's Theorem

In 1948, Claude Shannon demonstrated that if random noise is present, the maximum number of signals that can be identified correctly is equal to the square root of (1+S/N) where S/N is the ratio of signal to noise. He also demonstrated the relationship between the number of signal types and the number of bits each signal can carry: X signal types allows log X (to the base 2) bits per signal. Although it isn't simple, this means that the ratio between the number of signal types and bits a signal can carry is expressed mathematically as:

$$\log_2(1+S/N)^{1/2} = \tfrac{1}{2} \log_2 (1+S/N)$$

It isn't important that we understand the mathematics for the purposes of this book, nor is it really important to understand the formulas. What is important is that we have an appreciation for the relationship between passband, bandwidth, carrying capacity of a channel, and noise. These have been scientifically proven and tested over the years, and provide some engineering insight into why the telephone network works the way it does. It also helps address some of the technical barriers to increasing bandwidth available over a voice channel, particularly that of a local loop designed to carry voice traffic.

Switching

In the next chapter, we'll delve into the *Internet protocol* (IP). To understand network layer protocols, it's important that we have a basic understanding of switching.

At the beginning of this chapter, we discussed the prohibitive nature of providing an any-to-any network with wired connections from every device to every other. Using this approach, even a simple network of 100 computers requires approximately 5,000 links to make all the connections. The problem is one of scalability; a network using this design works fine to connect a few computers, but the design doesn't scale to the size necessary for today's expanded network environments.

An alternative approach is to connect devices with some form of switching. *Switching* is the process of placing temporary links between devices to allow communications. To be effective, these switches must be capable of routing around failed links and of providing some mechanism for congestion control when areas of the network are busy. Network layer protocols, like IP, address these issues.

Two different types of switching are in use in today's networks—circuit switching and packet switching or store-and-forward switching.

Circuit Switching

As the name implies, circuit switching is a technique whereby the switches establish a dedicated electrical (or optical) path between devices. The important point here is that a path or circuit is established for the duration of the call and is dedicated to the call. These resources in the network cannot be used for other calls or by any other user until the call is completed and the resources are released and available. In the telephone network, the circuits are set up before the call is connected, then released after the parties hang up the telephone.

Because networks cannot be designed to support every possible telephone call at the same time, these switches are designed as *blocking switches*: when all available resources are in use, callers will experience *queuing delay* or blockage until resources become available.

The delay through the network once a connection is complete is minimal. Most of the telephone network has been designed to provide about 55 milliseconds of delay over established circuits.

The two most common circuit-switched networks in use today are the public switched telephone network (PSTN) and the *circuit-switched public data network* (CSPDN). In the United States, services referred to as "Switched 56" are an example of the latter. With the advances of optical networking technology during the past few years, there is also considerable speculation that switching optical circuits may become a common practice.

Circuit switching can be implemented in several different ways. Space, time, and frequency each might provide the dedicated resource used in providing the path. *Space division multiplexing* is the oldest form used and provides a spatially unique path through the CO switch. In their earliest implementations, these switches were manual switchboards used in telephone offices. Operators manually plugged in cords to provide the necessary circuit connections. *Step-by-step* and *crossbar switches* are also space division switching techniques. Newer switches use computer programs to control electronic components.

Time division switching dedicates a time slot to the parties involved in a call. The two parties are guaranteed a *time slot*, or all of the bandwidth for a guaranteed portion of the time. In this method, the equipment used at each end knows what time slot to use, and both ends communicate during the assigned time slot. Because these time slots may occur several thousand times per second, the human ear cannot detect the underlying technology. A more complex variation of this approach is used in many CO switches and private branch exchange (PBX) systems today.

Cellular telephone networks and cable modem systems use an approach called *frequency division switching*. In this approach, communicating parties are assigned a dedicated slice of the passband within the total bandwidth available in the communications channel.

The key to each of these approaches is that something is dedicated to the call for the duration of the call. This dedicated resource cannot be used by any other user on the network until the call completes and the resource is returned to a pool of available resources.

Because circuit switching guarantees a dedicated path that cannot be shared, it requires a significant engineering effort to locate, reserve, and connect the necessary resources through the network. This drives the cost of a connection up and causes some delay in the setup process. As a result, circuit switching is more economical for connections of a longer duration, such as a voice telephone call, where parties may talk for three or four minutes. It works best when network utilization is high, providing a usage level that keeps the resources busy, but not overloaded.

Packet (Store-and-Forward) Switching

Many ways exist to provide switching without the use of dedicated facilities. One example of a store-and-forward switching network is the subway system in New York. Passengers can travel between any of the subway stations along the route. The topology of this network is referred to as a *hub and spoke* topology. The subway has many switching points or *nodes*. To get from one location to another, users might have to transfer from one line to another at one of these nodes. At the hub nodes, passengers (the traffic) might have to wait in a *buffer* (be stored) until the next available train arrives so that they can move (be forwarded) on to their destination. Just as passengers encounter delays in waiting for a train to arrive, and sometimes queuing delays when the trains are fully loaded, a store-and-forward network provides service that has very different characteristics from a circuit-switched network, which would be more similar to a New York taxi—dedicated to a passenger for the duration of the trip.

In a data network, even the links between the switches are shared on demand. Switches perform routing calculations to determine which link to send the data on to, then the data is placed in queue for that link. Resources are allocated on a first-come, first-served basis, and there are no guarantees that the next leg of the path will be available upon arrival. Delays in queuing can cause data to sit in buffers. In a data network, this means that the delay through the network can be sporadic and unpredictable.

Because of this unpredictability, large blocks of information aren't well suited to this type of network. Large blocks of information must be broken into smaller chunks in order not to degrade performance of the network. When we think of a four-minute telephone call, we are really thinking of a very large block of data.

In a store-and-forward network, each block of data has to carry some form of addressing information that the switches can use to determine its the final destination. Without this, the information can never be delivered to the recipient.

Data applications are often described as being "bursty in nature," meaning that there may be lapses or pauses between transmissions. Unlike a voice call, which is a real-time interaction between two people, a data connection is often an interaction between two computers, without a person directly involved. Because store-and-forward or packet networks use *statistical multiplexing,* or *first in, first out* (FIFO) methods, this type of network is better suited to a bursty type of traffic, like data.

Packet switching is the most common form of store-and-forward switching in use today, with *routers* being a perfect example of a store-and-forward switch. Packet switching breaks blocks of information into a predefined size or size range. This process of packetization does create some overhead, because each packet must have addressing information. Error checking can be performed on a per-packet basis and, if errors occur, only the corrupted packet must be retransmitted. This gains some efficiency in the network, because long messages do not need to be repeated entirely if an error occurs.

Connectionless versus Connection-oriented Networks

The circuit-switched network is clearly a connection-oriented network: the connection is the call setup process that establishes the circuit.

Packet networks can be either connection oriented or connectionless. In a connectionless network, no setup is required. Each packet carries sufficient addressing or routing information to allow it to be passed from node to node through the network. Because there are no guarantees or dedicated network resources, these networks are referred to as "best efforts" networks, and performance may be unpredictable. In cases of network congestion, these networks often discard packets to alleviate congestion or crowded buffers.

Connectionless networks also do not guarantee that packets will be delivered in the order they were transmitted. Packets might take different paths through the network and arrive at different times, so that the device at the recipient must have resources to store the packets until enough have arrived to reassemble the message for delivery.

The postal network is a good analogy of a packet network with no guarantees. You could receive this book in many packets, each containing a page. The envelopes would be the packets containing the message. At the receiver's end, the pages would have to be stored until all were received, then the book could be assembled. If a packet were damaged along the way, only the damaged page would require retransmission, not the entire book.

As you can see, packet networks provide a good technology for delivering short or bursty messages that don't require the overhead of call setup. Transaction processes, like ATM machines or credit card verification terminals, generate short messages that are ideally suited to this type of network.

In the next chapter, we'll look at the Internet protocol (IP) as a specific packet network and begin to delve into how voice telephone calls can be carried using a packet technology.

The Internet and IP Networking

To understand the basics of IP, we must review some general networking concepts to ensure that we have a solid foundation to build upon.

OSI Network Reference Model

Computer systems consist of several components, all necessary to communications, but we will not delve into the internal operations of computers—rather, we'll stick to basic components. At the most basic level we have the user, hardware, and software. Software falls into two distinct categories: applications and operating systems.

Operating systems are software that controls the basic functions of the computer. It handles memory partitioning, task assignment, and manages the flow of information to and from the *central processing unit* (CPU). Windows 98, Windows 2000, Linux, OS 8.1, and Cisco's IOS are examples of operating systems. Applications are programs designed to run under the operating systems and carry out some useful task. Applications include word processing programs, spreadsheets, graphics programs, games, and other useful utilities. They are designed to run underneath a specific operating system.

Because data communications is a very complex process, it has to be broken down into manageable sections, or layers. Each layer provides for some portion of the tasks involved in the network. One of the mostly widely used approaches to this layered concept is the OSI Reference Model, shown in Figure 2.1. The greatest benefit to a reference model is that it provides a common language for developers to use in discussing networking. Networking is a huge and daunting task, not unlike eating an elephant. The OSI model gives us a structure so that we don't have to eat the elephant in one bite, but can break it into manageable sized pieces.

Figure 2.1
OSI reference model.

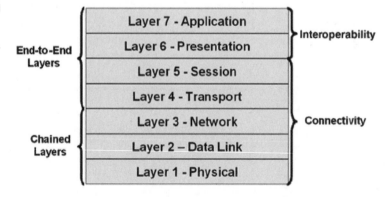

The Open Systems Interconnect (OSI) Reference Model serves two purposes:

- It provides a framework for dealing with connectivity and interoperability between dissimilar systems.
- It provides places for different protocols that perform different functions.

The layers are defined in two categories. The *end-to-end layers* are primarily focused on the total communication from end system to end system, or user to user. The *chained layers* are present in every node of the network, whether a personal computer or a router, along the path data might travel. These layers provide a chain of connections that allow data packets to be passed from node to node over a variety of disparate network architectures.

The Physical Layer

The Physical Layer describes a set of procedures or operating rules that provide for the mechanical, electrical, or optical transmission of bits along a physical link. This layer defines physical connections like DTE, DCE, and V.35.

This transmission can be *simplex* (one-way), *half-duplex* (alternating two-way), or *full duplex* (two-way). It can also be a *serial transmission* of a single bit at a time, or *parallel transmission* of several bits over several parallel electrical connections. Parallel connections can be expensive, and are best used for short distances, like the ribbon cable connecting components inside a computer.

Physical layer signals are sometimes *baseband*, or digital signals, and sometimes *broadband* or analog signals.

In the previous chapter, we reviewed theories by Harry Nyquist and Claude Shannon. At the physical layer, these theories are applied. Nyquist's theorem determined that the maximum signaling rate for any bandwidth can be determined by calculating 2× maximum frequency of the bandwidth. Shannon's Law states that bandwidth and the signal-to-noise ratio limit the information rate over a communication channel. In an electrical circuit, distortion caused by attenuation is fixed over distance. Noise, on the other hand is random. It is an unwanted signal that adversely affects the transmission. To provide clear communication, we must engineer the channel to expect both distortion and noise.

The Data Link Layer

The Data Link Layer is responsible for providing error free transmission from node to node over a physical medium that is prone to errors. It only provides error detection across a single physical link. To accomplish this, the data is *framed*, or structured, and information is added to allow each node to determine whether what was received matches what was transmitted. This is easily accomplished by performing a *cyclic redundancy check* (CRC) against received data frames.

Basic functions performed at the Data Link Layer are:

- Framing, or delineating the beginning and end of the data,
- Establishing a control mechanism for access to the media, like CSMA-CD in an Ethernet LAN,
- Detecting errors, usually by a CRC, and
- Error correction in some cases, typically in a connection-oriented environment. Connectionless networks generally discard errored data.

Some examples of Data Link Layer technologies are Ethernet, Token Ring, Frame Relay, and asynchronous transfer mode (ATM).

The Network Layer

The Network Layer is used to transport packets across the network using some form of routing, provide for congestion control, establish an addressing scheme, provide either connection-oriented or connectionless services, and define an addressing scheme. In a packet environment, the Network Layer protocol must define the structure of a packet.

Some common Network Layer protocols in use today are:

- Internet Protocol (IP) in the TCP/IP protocol suite
- Internet Packet Exchange (IPX) in Novell Netware
- The Path Control layer in IBM's Systems Network Architecture (SNA)

Packets are transported from one end of the network to the other by passing from node to node.

Congestion control might be handled by changing the size of buffers in various devices along the path. In many networks today, congestion is dealt with by simply discarding the packet. Higher-layer protocols are robust; they detect the nondelivery of packets and automatically request retransmission in many cases.

Routing protocols are the subject of numerous books, and not something we will go into here in depth. A routing protocol is just a set of rules that the nodes follow. It's a procedure used by the network to determine how to transmit a packet from node to node. A routing protocol might be centralized, but a central routing database is impractical in a network like the Internet, so we use decentralized routing protocols, with some information being contained in each router.

One very crucial aspect to routing protocols is that they must be dynamic, or sensitive to changes in the network. The Internet has many networks and hosts connected. These connections come and go as links fail, as new networks are added, and as changes occur. The network, or the nodes within the network, must learn very quickly about changes in network topology, so that packets can be delivered from end to end.

The Network Layer protocol is where the addressing structure for a particular network is defined. IP has a specific addressing scheme, but IPX in Novell networks uses a different scheme, and DDP in the AppleTalk environment yet another. The Network Layer protocol defines the addressing mechanism required for each specific protocol.

Because a packet network doesn't establish a circuit or connection, the packet itself must contain both a source and destination address. The analogy of a letter mailed at the post office is frequently used. It has both the recipient's address for delivery, and the sender's address for any return message, such as notification of nondelivery.

The Transport Layer

The Transport Layer is responsible for error control and often flow control, on an end-to-end basis across the entire network. This is different from error control at the Data Link Layer, which is only from one node to the next. The Transport Layer provides a mechanism for recovery from errors in the network.

Transport Layer protocols that require a high degree of reliability are more complex, whereas simple protocols provide less-reliable service, but with less overhead. We'll see this comparison directly in Transmission Control Protocol (TCP) and User Datagram Protocol (UDP), which are both used in delivery of Voice over IP (VoIP) packets.

The Session and Presentation Layers

The Session Layer controls the management of individual user sessions, logging on and off, and recovery from some failures.

The Presentation Layer deals primarily with the form and syntax of the information being transmitted and passed to the Application Layer. In human terms, this is analogous to translating between different languages, which might use different alphabets and different phonetic intonations. Code conversion takes place in this layer.

We won't be exploring either of these layers in any detail here.

The Application Layer

The Application Layer provides a set of protocols that deal with the meaning and form of the information. These protocols provide the syntax for a task to be completed. Some examples of Application Layer protocols include:

- X.400
- Domain Name Service (DNS)
- Simple Mail Transfer Protocol (SMTP)
- File Transfer Protocol (FTP)

Earlier, we mentioned two types of software: applications and operating systems. It's worth remembering that the operating system in a piece of hardware is an application, and most have some functions at the Application Layer. An operating system can also operate at other layers of the reference model, depending on its construction and design.

The Importance of TCP/IP and the Internet Protocol Suite

In the early 1960s, a militia group blew up four microwave towers in Utah. This event cut off communications in the western United States. The circuit-switched PSTN was a part of our daily lives, and we realized it was vulnerable. It was also expensive to build equipment and components for this network. Communications are vital, and the Department of Defense recognized a need for other means of communication. The Advanced Research Projects Agency (ARPA) was formed, and research began into the idea of packet networks. The first node on the ARPANET network that followed was installed in 1969 in Los Angeles.

By the 1980s, the corporate world had embraced the computer as a business tool. Companies invested deeply in systems from IBM, Honeywell, Sperry, and others. As the business climate changed, we saw growth, mergers, and acquisitions occur on a regular basis. Companies had to find a way to integrate disparate computer systems from different vendors into a single environment.

Companies began using the TCP/IP or Internet model that had continued to evolve from the early ARPANET work. TCP/IP was, and is, an open standard, well documented and defined, to which any vendor or manufacturer can comply. It provides a platform for multivendor interoperability that had not previously existed. The Internet gave rise to *intranets* (internal networks within a company or enterprise) and *extranets* (external networks linking to partners, customers, and other organizations). Today, we hear the term *enterprise network*, which typically involves the Internet, intranets, and extranets all in combination to conduct business. The migration to TCP/IP in the 1980s was the beginning of the process of *convergence*: the unification of applications and services onto a single network and single device.

The TCP/IP Suite

When someone says they are running IP, they mean far more than just the Internet protocol at the Network Layer. IP refers to the TCP/IP suite, a collection of protocols and services that are widely used today. Figure 2.2 provides a summary of how the components of the TCP/IP suite relate to each other.

Note that the TCP/IP approach is often described as a four-layer model, rather than the seven-layer OSI model we discussed earlier. The correlations between the two models are noted on the figure. Both are used somewhat interchangeably, to the confusion of many. For our purposes, we'll focus on the Network Layer (called the Internet layer in TCP/IP jargon), the Transport Layer, and Application Services.

Several protocols operate at the Network Layer. IP provides a packet service that is connectionless in nature, using no predefined circuits or connections. The primary function of IP is to deliver packets, but it cannot do so without the help of other protocols. Without these other protocols, IP would not be able to function.

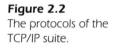

Figure 2.2
The protocols of the TCP/IP suite.

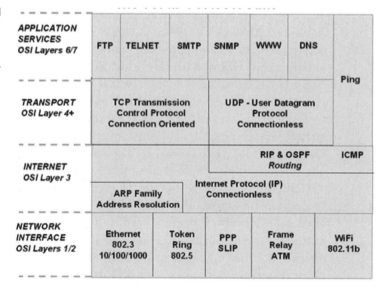

The *Address Resolution Protocol* (ARP) provides a correlation between the hardware address of interfaces at the lower levels and the network address used for routing packets. ARP is used to correlate a known IP address with the Media Access Control (MAC) address at Layer 2. Reverse ARP (RARP) resolves a known Layer 2 address to the network or IP address. Other variations have been developed to support technologies like Frame Relay and ATM.

The routing protocols provide information about paths through the network that IP can use to deliver packets. Routing protocols, such as Routing Information Protocol (RIP) and Open Shortest Path First (OSPF), often operate directly at the Network Layer, and may also use Transport Layer protocols to communicate.

The Internet Control Message Protocol (ICMP) provides management functions in the network. It's mostly commonly used, as shown, to *ping* a device to confirm reachability. ICMP also provides error messages about packets or unreachable devices. ICMP doesn't use TCP or UDP at the Transport Layer, but sends information directly inside IP packets.

Internet Protocol (IP)

IP was introduced in the early 1980s, and is widely used in conjunction with TCP at the Transport Layer. It is the protocol used in the Internet and has become the de facto standard for data communications.

Two compelling reasons exist for the popularity of IP: it has a well-documented history of implementation, and it is an open standard in the public domain. No one company owns or licenses IP. The community of interest has jointly developed it over many years. This means that anyone can implement IP, in hardware or software, without paying licensing fees.

IP provides a network service that is described as best effort or unreliable, and connectionless. Messages are broken up into packets and transmitted individually. Every IP packet contains information about the source and destination addresses, and these packets can, and often do, take different routes to be delivered. A user transmits a packet into the network with enough information so that the network can deliver it.

IP is referred to as an *unreliable protocol* because there are no guarantees of delivery from IP itself at the Network Layer. There are also no guarantees that packets will be delivered in the same sequence they were transmitted. Just as the post office doesn't guarantee delivery of a letter, IP doesn't guarantee delivery of a packet. The network makes its best effort to deliver the packet, but without any guarantee. That isn't to say that delivery cannot be assured. In the postal system, guaranteed delivery is assured by paying more and sending a letter via registered mail. In an IP network, the same applies. To guarantee delivery and sequentiality, we must pay more, in the overhead of some additional mechanism like TCP. We'll dig deeper into this later in the chapter.

Routing protocols, as we mentioned earlier, provide information about the various paths available through the network. They build routing tables, which IP consults to determine the best path for delivery of a packet at the time the packet is transferred. As routes fail or are added, the routing table changes. The best path varies due to changes in the topology of the network. This table look-up, while potentially overhead intensive, provides the *best efforts* aspect of IP.

As a Network Layer protocol, IP must also provide a method for dealing with congestion in the network. Network traffic is unpredictable and, in peak periods, there may not be enough resources to deliver every packet. IP discards packets when congestion occurs. A variety of approaches to packet discard exists, and we'll review those further when we discuss *quality of service* (QoS) later in the book.

Because IP is a connectionless protocol, each packet must be processed individually at each node along the transmission path. It is a form of packet, or store-and-forward, switching as discussed in the first chapter. See Figure 2.3.

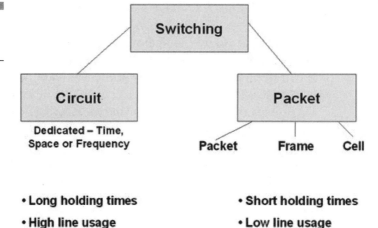

- **Long holding times**
- **High line usage**
- **Sensitive to Delay**

- **Short holding times**
- **Low line usage**
- **Tolerant of Delay**

The IP Packet. As a Network Layer protocol, one of the things IP must do is provide a packet structure to gather enough information to route the data packet by packet across the network. IP defines mechanisms for breaking large segments into smaller ones, which is sometimes necessary when transmitting over different technologies. The packet structure also defines methods for controlling the lifetime of a packet in the network and for checking the integrity of the header for errors.

This is a good time to point out that, as a continually evolving open standard, the version of IP used is a factor. Because version 4 is the universally accepted standard in use in the Internet, that's the version we'll review. IPv6 is being tested and implemented by some, but has not yet become universal in deployment.

The IP packet consists of two portions: the header and the data. The general structure of an IPv4 packet is shown in Figure 2.4.

Ver	IHL	Type of Service				Total Length	
Identification			X	M F	D F	Fragment Offset	
Time to Live		Protocol				Header Checksum	
Source Address							
Destination Address							
Options						Padding	
Data/User Payload							

- The Version field is 4 bits long and contains the IP version number in use. This is set to 4 (binary 0100) in today's IP networks.
- The Internet Header Length (IHL) field is 4 bits. This field identifies the length of the header in 32-bit words, each of 4 octets. IP supports options that allow the header length to vary, but this is normally set to 5 (binary 0101) because the standard header is 20 octets in length. This field is necessary to locate the boundary where the header ends and user data begins.
- The Type of Service field provides a prioritization mechanism that has rarely been used in IP. It allows the originating station to request a class of service from the network. We review this field a bit more in the chapter discussing quality of service (QoS).
- The Total Length field is 2 octets long. These 16 bits give the total length of the packet in octets. An IP packet can be as large as 65,535 octets in length.
- The Identification field is used to identify individual and unique packets in the network, but most importantly in conjunction with *fragmentation*. When a packet is fragmented, this field, coupled with information in the next field, provides information that allows the original packet to be reassembled at the receiver. A combination of source address and identification can uniquely identify every packet in the network.
- Fragment Offset, More Fragments, and Do Not Fragment fields are used to provide a sequencing mechanism for breaking large packets into smaller packets for transmission and reassembly at the receiver. More Fragments and Do Not Fragment are 1-bit flags used in this process.
- Time To Live is set by the originating station and is decremented at each node along the path. It uses hop count as a mechanism to control how long a packet can exist in the network. When TTL=0, the packet is discarded so that if a packet is caught in a loop, it does not continue to exist forever in the network.
- The Protocol field indicates which protocol is being carried inside the IP packet. Because TCP, UDP, ICMP, OSPF, and other protocols transmit data directly in the IP packet, this field provides an identifier for the packet contents.
- The Header Checksum is recalculated at each node as the packet moves hop by hop through the network. It is used to verify the integrity of the header.
- Source Address and Destination Address are 32-bit fields to identify the sender and receiver. The source address is necessary not only for

identification, but for responses and any attempts at nondelivery notification.

- The Options field may or may not be used. If used, it follows immediately after the standard 20-octet header. This field is of variable length, up to 40 octets, and is made up of several smaller fields. We won't explore it further.

- *Padding* is used in conjunction with the Options field to provide 32-bit alignment of the information in the header and options fields.

IP Addresses

Every device connected to an IP network must have a unique address, which is used to deliver packets of information. The IP protocol defines this address as being 32 bits of binary information.

For convenience, we've broken the binary information in the IP environment into *octets*, groups of 8-bit increments. In the addressing environment, dealing with a 32-bit binary number is very easy for computers, but very difficult for people. Thirty-two bits of address space creates the possibility of over 4 billion unique addresses that can be assigned to hosts connected to the network. To ease management of addresses, we use a format referred to as *dotted decimal*, where octets are represented by decimal numbers that we're familiar with, and separated by a decimal point. Figure 2.5 demonstrates the basic conversion of these 8 bits.

Figure 2.5
Binary to decimal conversion of IP address.

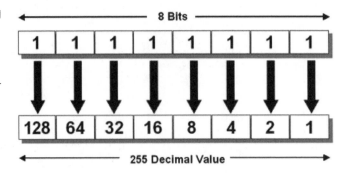

We see that an octet can range from a value of 1 to 255 in decimal, based on every possible combination of individual bits. So the IP address 192.168.1.50 is a decimal representation of the binary address 11000000 10101000 00000001 00110010.

IP addresses are hierarchical and are divide into two components: the NETID and HOSTID, similar to the telephone number hierarchy in the United States. Telephone numbers are represented by a three-digit area code, a three-digit prefix, and a four-digit station number. The NETID identifies the network the host is attached to, and every host on a given network will have the same NETID. Conversely, the HOSTID is unique to the device within a network and no two devices with the same NETID can share the same HOSTID. The NETID is of particular importance because routers use it to deliver packets. The HOSTID only has local significance after a packet has reached a destination network. Figure 2.6 shows the relationship of the HOSTID and NETID.

Figure 2.6
NETID and HOSTID.

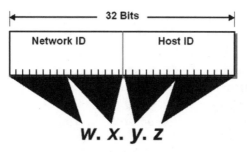

The NETID and HOSTID cannot be made up of all ones or all zeros. In an IP network, all ones in an address represent a *broadcast* that is transmitted to all devices on the network. All zeros represent *this host*, this machine.

The determination of which portion of the IP address comprises the NETID and which portion comprises the HOSTID is not as clear-cut as Figure 2.6 might suggest. Network addresses are also divided into *classes* for different sizes of networks. The first octet of the address is the key to identifying the class of network in use.

If the first octet is from 1 to 126, it is a Class A network address. It uses the first octet for the NETID, and the remaining 3 octets for the host ID. This leaves 24 bits of address space for host, which means we can have a network of approximately 16 million hosts, but only 126 of these networks. Clearly, not many businesses require this class of address.

If the first octet is from 128 to 191, it is a Class B network address. It uses the first two octets for the NETID and the remaining two octets for the HOSTID. This 16-bit address space means we can have approximately

64,000 hosts on these networks. Still, many companies are much smaller and do not require so many addresses.

If the first octet is from 192 to 223, it is a Class C network address. In a Class C network, the first three octets are used to represent the NETID, and only the last octet is used for the HOSTID. This only allows 254 hosts per network, but is a widely used addressing scheme.

We didn't mention 127 or 224 through 255 in the first octet because these are reserved for other purposes and not used in operational networks.

Some interesting mathematical tidbits are hidden here, which make this an elegant addressing scheme. The 32-bit address is divided into 4 octets, and earlier we looked at the value of each bit. If the very first bit in an address is set to 0, this address must be a Class A address. We know by looking at a single bit that the first octet represents the NETID and the remainder the HOSTID. That first bit has a value of 128. If we set it to a 1, the value of the octet has to be 128 or greater. If the first two bits are 10, we know that the address is a Class B address, because if both these bits are set to 1, the value of the first octet must be 192 or greater. This approach allows a computer, by looking at a single bit in the address, to determine where to "draw the line" between the NETID and HOSTID.

Other approaches to addressing exist, and we now often use what is termed *classless addressing* to get networks of different sizes. We won't discuss these approaches in depth here, but it's important to understand that there is a very important distinction between the NETID and HOSTID. To deliver packets across the network, routers must be able to identify the NETID.

Again, the post office analogy is a great comparison to the IP network. In Figure 2.7, we have three houses existing on Main Street. A letter addressed to the resident at 2 Main Street works its way through the postal system. The postal system handlers don't need the house number of the addressee. They only need to know how to get the mail to the carrier for Main Street. Once it reaches the carrier on the Main Street route, there is only one house number 2. If there were duplicate numbers, the mail couldn't be delivered. There may be a number 2 on other streets, but 2 Main Street uniquely identifies this house.

The IP network is not much different. A packet addressed to 192.168.1.50 is transmitted into the network from 177.115.23.8. The network, like the postal system, doesn't need to identify the specific host, but merely the network, analogous to the street, on which the host resides. The packet is passed from router to router along the network until it reaches the router for the local network.

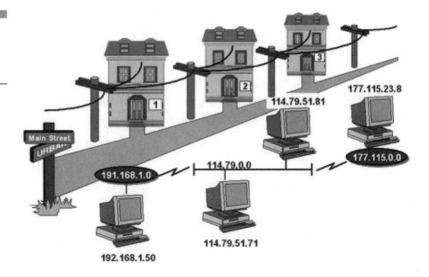

Figure 2.7
The IP network is similar to the postal network.

Acquiring an IP Address

Anyone on a private network can use an addressing scheme of their own choosing. In the early days of local area networking, this was a common practice. The problem came into focus as networks began to connect to the Internet. If an administrator built a private network, then later attempted to connect to the Internet, they often found the address already in use. To connect to the Internet, a system administrator must have a *public IP address*, which means there has to be some form of address registration.

A number of addresses are reserved for private networks. These addresses really should be used within private networks for a variety of reasons, but doing so can also create other issues. We'll discuss one issue, *network address translation* (NAT), in an upcoming section. These *reserved addresses* are documented in RFC 1918 from the Internet Engineering Task Force (IETF–www.ietf.org):

- 10.0.0.0 through 10.255.255.255
- 172.16.0.0 through 172.31.255.255
- 192.168.0.0 through 192.168.255.255

IP addresses have historically been managed by the Internet Assigned Number Authority (IANA—www.iana.org), but during the past two years have been in a state of transition to the new Internet

Corporation for Assigned Names and Numbers (ICANN—www.icann.org). This organization, with the help of registry partners, assigns blocks of NETIDs to geographic areas. In the past, there were three registries: American Registry for Internet Numbers (ARIN) serving the Americas; Asian-Pacific Network Information Center (APNIC) serving Asia, Australia, and the Pacific; and Réseaux IP Européen (RIPE) serving Europe, the Middle East, and parts of Asia and Africa.

With the growth of the Internet during the past several years, and the transition to ICANN as the authority responsible for assigning addresses (and domain names), we now have many registries to choose from. A current list is online at http://www.icann.org/registrars/accredited-list.html.

In the business world today, *Internet service providers* (ISPs) are assigned blocks of numbers, which they then assign to their customers, although some larger companies do still obtain dedicated IP addresses directly from ICANN. The service provider assigns end networks a NETID from the ICANN-assigned block. The system administrator then assigns local HOSTIDs within that block to individual hosts on the network, as shown in Figure 2.8.

Figure 2.8
Network ID assigned by ICANN; Host ID assigned locally by system administrator.

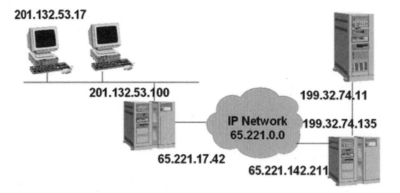

201.132.53.17

201.132.53.100

IP Network
65.221.0.0

199.32.74.11

199.32.74.135

65.221.17.42

65.221.142.211

Dynamic Addressing

The IP address of a host, whether the host is a personal computer, a router, or some other device, must be programmed into the memory of the host. The technique of assigning a permanent address to a host is called *static addressing*. Although this was common practice in the past, more and more system administrators lean toward *dynamic addressing* in today's networks.

The *Dynamic Host Configuration Protocol* (DHCP) provides a means for a host to automatically request an address when needed, then to release that address back to a pool of available dynamic addresses when it is no longer required. DHCP is based on an older *Bootstrap Protocol* (BOOTP) that was widely used for static addressing in the past.

Dynamic addressing is ideally suited to an ISP with dial-up users. Users log on to the network and need an address while they check email and surf the Web. But once they have finished, that address could be used by another user dialing in to the network. Dynamic addressing provides a mechanism for the automatic reassignment of unused addresses. With the surge in popularity of laptop computers and mobile users, it hasn't taken long for DHCP to become necessary for many business networks as well.

DHCP has another benefit for the system administrator. Because the address can be assigned automatically when the user connects to the network, a properly configured computer only needs to be set up once. If there is a change in provider or addressing structure due to some design change in the network, each user's computer does not have to be manually reprogrammed with a new IP address. Given the intense labor effort this has sometimes required in businesses, system administrators have been actively converting many networks to use DHCP for addressing.

Dynamic addresses have also become an important issue for DSL and cable modem users. The idea of a high-speed connection to the network is very appealing. However, there is a potential danger to the "always-on" PC connected to the Internet. The danger is that hackers might attempt to gain access to the user's computer for some ill-intentioned purpose. If the victim's IP address is static and never changes, the hacker will always be able to locate the computer.

Cable modem providers initially used static addressing in many cases, but quickly discovered another problem. Although the computer might be "always on," it wasn't always in use. Inactive computers were consuming addresses, and the cable modem providers were rapidly running into a lack of addresses. In a community of 1,000 subscribers, it has been somewhat uncommon to find every user trying to use the network at the same time.

Dynamic addressing solved both these potential problems, and it is generally simpler to manage than a static addressing scheme. DHCP is a widely accepted protocol by which host systems request the temporary lease of an address when they connect to the network. At the end of that lease, a system that is actively being used on the network will renegotiate either the extension of the lease, or a new lease for an address. Idle systems will simply return their address to the available pool, and the lease expires.

Dynamic addressing has proved beneficial in many different areas, but there are still situations where it doesn't fit. Servers and routers are not suitable candidates for DHCP and dynamic addressing. Because these devices provide services to either users or to the network, they must be reachable and easy to find. Changing addresses on these devices requires a system for sharing the addresses. The *Domain Name Service* (DNS) is used for domain and server names, but it could not keep up with constant changes of addresses in a network as large as the Internet.

As we move into the IP telephony aspect of the network, logistics becomes important. On an IP network, a host's IP address is how it is reached. To place an IP phone call to a host, the address is necessary. Imagine the difficulty of placing telephone calls if telephone numbers were not static, and every time you tried to call home, you had to look up a new telephone number. Dynamic addressing is an area that is key to the implementation of IP telephony and could cause problems if mismanaged. Work is in progress in the IETF to incorporate dynamic updates into DNS and to do so securely.

Private Addresses and the Public Network

Earlier, we mentioned some ranges of addresses that are suitable and recommended for use within the private network. The use of private addresses gives a great deal of freedom to the network administrator designing an addressing scheme. It can also provide a greater level of security, because private IP addresses are never seen on the public network.

This security also creates a problem: if a private address cannot be transmitted on the public network, private addresses must be translated into public addresses, and then back to private addresses in many cases. This requires a special gateway device capable of performing network address translation (NAT). A NAT device provides a method of correlating internal private addresses to external public addresses. It essentially acts as an addressing "proxy" for the private network. The NAT device needs one or more public addresses, and the system takes packets from the private network, *translates* the address to one of its public addresses, and maintains a table of these correlations to ensure responses get directed to the appropriate IP address on the private network when they come back. NAT is a dynamic process and performed *on the fly* as packets are transmitted.

Another benefit of NAT is that fewer public addresses need to be assigned to the network, leaving the ISP with addresses that can be assigned to other users and businesses. NAT is an excellent mechanism for conserving address space. IPv4 has the capacity to address approximately 4 billion hosts, but due to inadequate planning and unanticipated growth in the Internet, address space is being exhausted and addresses are in short supply. Using private address space in the private network, in conjunction with NAT, alleviates some of this problem and could conceivably provide addressing for 4 billion networks rather than 4 billion hosts. Converting to a newer version of IP will be an arduous task, and extending the usable lifetime of the current widely deployed version is a benefit to all users.

Reliability versus Quick Delivery

We've described IP thus far as being an unreliable protocol without guarantees, but what happens when an application requires delivery? What if the application doesn't require any guarantee of delivery, but needs to deliver the information as quickly as possible? IP provides *best efforts* service, with no guarantee.

We have to implement a technique for assuring the type of delivery by adding another protocol, but we do this at the Transport Layer. Referring back to Figure 2.2, the TCP/IP suite, we must now delve into the Transmission Control Protocol (TCP) and User Datagram Protocol (UDP). When we add another protocol, we begin to have a measurable affect on the data and must consider what happens when we transmit data.

In our discussion of the IP packet, we didn't investigate the User Data field in detail. User data is that information being transmitted from one system to another. Where does user data come from? The user generates it, of course, but it's created in an application of some kind, running at Layer 7 as an application service. To transmit data, we need to move it through the TCP/IP stack, which is a practical implementation of the OSI Reference Model. How this occurs is shown in Figure 2.9.

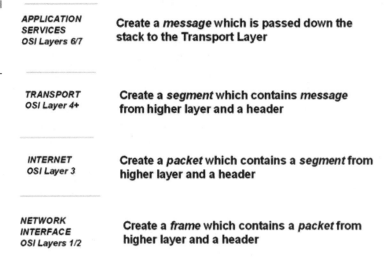

Figure 2.9
Data through the
layers of the OSI
model.

Notice that, at each layer, the data is placed into the payload of the lower layer and another header of some type is added. Each header at each layer introduces some overhead to the process, and all serve a purpose, but it is overhead nonetheless. The more overhead consumed, the less room there is for user data, and the real purpose of the network is to transmit user data.

The Transmission Control Protocol (TCP) is a Transport Layer protocol used to guarantee both delivery and sequentiality of packets transmitted over an IP network. IP guarantees nothing, so we must add the necessary overhead to provide this assurance. Recall the analogy of the registered letter we used earlier and the additional expense incurred to guarantee delivery through the postal system. TCP also has a cost, and that cost comes in the form of overhead added to the data stream.

Addressing the Layers

We've alluded to addressing at other layers in the model. This is a good place to clarify addressing, which exists in some form at many points throughout networking. At the Physical Layer, an address equates to either the wire itself, or a physical port on a hardware device. The Data Link Layer address is often referred to as the *MAC address*, in relation to the Media Access Control scheme deployed in the local area network (LAN) environment. We've reviewed IP addresses at the Network Layer in some depth.

The Transport Layer also has an address that consists of two parts. One part of the address is the IP address, but when you think about it, an IP address gets you connected to a computerized device. In today's networks, computers can have multiple processes running concurrently, and often do. One of the purposes of the Windows™ family of operating software is that users can open a "window" to more than one application and have multiple programs running simultaneously. In the networked computing environment, we need a mechanism to connect not just to the host, but also to a specific application running on that host processor.

Take for example, the server in Figure 2.10, with an IP address of 212.220.10.1. To verify the computer is connected and can be reached, we could "ping" it and get a response so that we know the machine is connected. But the machine is providing three separate services at the same time. It's a Web host, an email server, and an FTP file server, all in one machine. When we send information to the server, we need to not only send it to the computer, but to the application running within the computer.

Figure 2.10
Transport Layer addressing in IP is a combination of the IP address and the port number associated with the application.

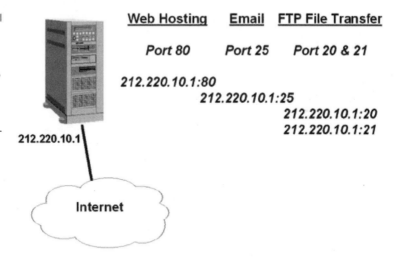

The *port numbers* shown are numbers associated with well-known applications, and these are used in combination with the IP address to reach a specific application within a machine. These numbers are not physical ports, but port numbers that have been assigned to certain applications as a default.

In Figure 2.10, a packet addressed to 212.220.10.1 will reach the machine. If the data is destined for port 80, as the information is delivered, it will be delivered to the web server application running within the server.

These ports fall into three different categories:

- Well-known ports are numbered 0 through 1023.
- Registered ports are numbered from 1024 through 49151.
- Dynamic or private ports, often referred to as the *ephemeral* ports are numbered from 49152 through 65535.

Assigned ports are all documented at www.iana.org/assignments/port-numbers. Some examples of *well-known port* assignments are listed in Table 2.1. Although this list is far from complete, it gives us a good idea of how we can multiplex more than one application onto a network-connected device and communicate with all running applications,

TABLE 2.1

Well-known Port Numbers

Service Name	UDP	TCP
Client/Server Communication		135
DHCP client		67
DHCP server		68
DNS Administration		139
DNS client to server lookup (varies)	53	53
LDAP		389
POP3		110
SMTP		25
NNTP		119
File shares name lookup	137	
File shares session		139
FTP		21
FTP-data		20
HTTP		80
HTTP-Secure Sockets Layer (SSL)		443

continued on next page

TABLE 2.1

Well-known Port
Numbers
(continued)

Service Name	UDP	TCP
IKE	500	
Microsoft Chat client to server		6667
Microsoft Chat server to server		6665
NetMeeting Audio Call Control		1731
NetMeeting H.323 call setup		1720
NetMeeting H.323 streaming RTP over UDP	Dynamic	
NetMeeting Internet Locator Server ILS		389
NetMeeting RTP audio stream	Dynamic	
NetMeeting T.120		1503
NetMeeting User Location Service		522
SNMP	161	
Telnet		23

NetMeeting and H.323 were included on the list above because they specifically relate to IP telephony. These port numbers and this addressing scheme are critical components when we endeavor to make a voice call to an IP device on a network.

Depending on the documentation you read, you may hear this address referred to as a *port*, a *socket*, an *endpoint*, and perhaps a *transport service access point*. All are correct, and all refer to the same thing: the address at the Transport Layer.

Wherever possible, both TCP and UDP use the same port number to access the same application.

Transmission Control Protocol (TCP)

The Transmission Control Protocol (TCP) was designed to provide a connection-oriented service that guarantees both delivery and sequentiality. TCP was necessary, because the lower Network Layer is unreliable, and some processes and programs require a delivery guarantee to operate correctly.

Three primary features characterize TCP services:

- **Connection-oriented communications**—TCP connections consist of three phases, much like a telephone call: a setup process, a sequence of requests, and acknowledgments. This establishes a connection between the two parties. The data transfer phase then occurs, analogous to the actual conversation on a telephone call. Finally the connection must be terminated. Just as hanging up the telephone signals the network that a phone call has finished, TCP sends signals to disconnect the connection.
- **Buffered transmission of data**—TCP implements buffering at both the sender and receiver. TCP is receiver oriented, in that the receiver sets the size of the data segments to be transferred, but both sides use buffering to maximize throughput across the connection. TCP attempts to keep the connection stream and the buffers full for maximum efficiency.
- **Full-duplex communication**—Data is transferred in both directions simultaneously over the TCP connection, just as if there were two parallel connections.

Just as we have frames at the Data Link Layer, and packets at the Network Layer, we transmit the data in a defined format called a *segment* at the Transport Layer. TCP treats data as a continuous stream of information, and uses these segments to provide the setup and teardown of the connections provided for use.

Like an IP packet, a TCP segment is divided into a head and data area. The structure is similar to that of an IP packet, with basic fields, some options, and the user data at the end of the segment. Like an IP packet, the basic header occupies 20 octets of data; however, these fields perform different functions. See Figure 2.11.

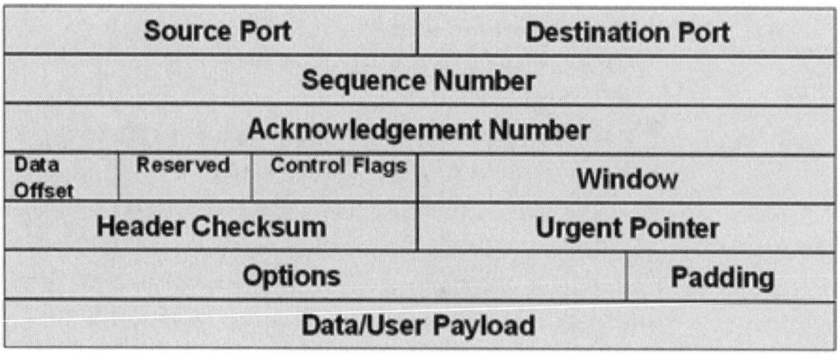

Figure 2.11
The TCP segment structure.

Source Port			Destination Port	
Sequence Number				
Acknowledgement Number				
Data Offset	Reserved	Control Flags	Window	
Header Checksum			Urgent Pointer	
Options				Padding
Data/User Payload				

Source and *destination port fields* are used to identify the port number associated with the application using TCP services. Notice that there is no IP address on the TCP segment, because there doesn't need to be. This segment becomes the *payload*, or user data in an IP packet. When it reaches the destination, the host machine receives the data based on the packet destination address. As the data is passed up through the TCP/IP stack, it then opens the TCP segment and reads the port information to identify which application process to deliver the information to.

Sequence numbers and *acknowledgment numbers* are crucial to TCP's capability of guaranteeing delivery and sequentiality. This mechanism provides both guarantees. The Sequence Number field is for sequencing the octets in the data stream. It signifies the position the octet occupies in the original stream of the message. This number is initialized at the time the connection is established.

The Acknowledgment Number field is used by the receiver to acknowledge data that has been successfully received. This field contains the sequence number that the receiver expects to see next. This provides a method for one acknowledgment to confirm receipt of several segments, which is how TCP is designed to operate. The sender transmits a series of numbered segments, for example 1,2,3,4,5,6,7. The receiver acknowledges this series by transmitting 8, the next segment expected. The sender knows that everything up through 7 has been successful received.

Control Flags is a 6-bit field used to send connection setup and teardown information, and to provide other control mechanisms, such as to look at an urgent pointer. The following options are available as control flags:

- SYN (Synchronize Sequence Number) is sent to establish a session. This flag begins communications and provides an indicator to initialize the Sequence Number field.
- ACK (Acknowledgment) is used to signify the validity of the Acknowledgment Number.
- URG is an Urgent Pointer and, if set to 1, provides instructions to read information in the Urgent Pointer field.
- PSH provides the Push Function, which can be used to move small bits of data ahead in the stream so they do not sit in buffers.
- FIN is the Finish command and is used for a normal, or graceful, close of a connection. Data in transit will still be delivered when the FIN command is executed.

- RST is used to Reset Connection. This provides a method for immediate disconnect, but unlike the FIN command, does not provide for delivery of information in transit. The abort is immediate and data can be discarded as buffers are emptied.

These flags are not only used for normal TCP sessions, but also provide a useful security function in many networks. A network device, typically a router or firewall, can inspect packets to see if the appropriate flag has been set, and discard packets which don't conform to expected operations. The SYN and ACK flags provide information about the direction of the traffic. If an internal host transmits a SYN, an ACK would be expected in the response. An inbound SYN without an ACK having been sent could indicate an unwanted attempt to access the network.

The *Window field* is provided for flow control purposes. We said earlier that TCP is receiver oriented and uses buffers. The Window field allows the communicating systems to share information about buffer capacity and the size of segments that can be transmitted successfully. This field allows for Window size to change during the course of the session. Using the Window field, the sender and receiver can fill the communications stream as completely as possible, while not sending data so fast that the sender overwhelms the receiver.

The *Checksum field* is used to check the integrity of the entire segment. It is a mechanism for ensuring data is received intact. If the Checksum fails, an acknowledgment (ACK) will not be sent for that particular segment.

Options and Pad fields may or may not be used. We won't review them here. They are often used to fine tune the performance of TCP, but should not be manipulated without a solid understanding of what reaction will follow.

Figure 2.12 illustrates how a TCP session is established, data is transferred, and normal completion. One important aspect of this communication is the *three-way handshake* that is necessary to guarantee delivery of the information. Remember that every TCP segment adds 20 octets of overhead to the IP packet payload.

Therefore, although TCP provides a reliable, connection-oriented service, it might also create an overhead-intensive situation, depending on the type of traffic. If Host A in Figure 2.12 needed to send the word "go," it would require eight TCP segments to complete the communication. Just as we noted earlier with circuit switching, a cost is associated with establishing connections. This cost may be perfectly acceptable for some types of traffic, but some traffic may not require the overhead associated with the assurances TCP provides.

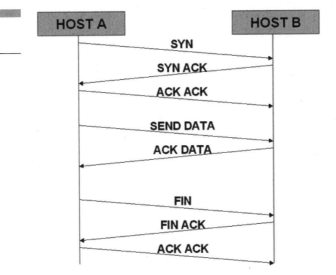

Figure 2.12
The TCP session.

Other performance issues surrounding TCP won't be discussed here. System management issues in the operating system may affect window size in TCP sessions, which can affect performance. Improper settings can induce delay. Buffer space and the ability of the operating system to perform context switching between active processes can create performance concerns. Even network reliability is a factor. TCP guarantees delivery, but a network that continually loses connectivity, or cycles between being up and down, will have a ripple effect in TCP. Segments may have to be retransmitted time and again in order to receive the acknowledgment and continue the transfer.

Given that TCP isn't suitable for all applications, particularly those that can't bear the price of the overhead, an alternate Transport Layer protocol must be used.

User Datagram Protocol (UDP)

The User Datagram Protocol (UDP) provides a service at the Transport Layer that is both connectionless and best efforts, just like IP. There is no sequencing of information and no acknowledgments are returned. Information is sent, and is assumed to be received at the other end unless some higher-layer protocol or application requests retransmission. This is not a flaw in design, but the purpose of the protocol. The protocol is simpler than TCP, which makes it easier to implement, and it reduces the overhead and potential performance degradation that TCP's higher bandwidth consumption can cause.

UDP was designed to support applications that are of a "request/response" nature; that is, a request is sent, a response is returned, and the transaction or session is completed. Some examples of this type of application have traditionally been the Simple Network Management Protocol (SNMP) and Domain Name Service (DNS). SMNP is used to set or get information in a "trap" for managing network devices. DNS is used to send a domain name to a server and receive an IP address as a response. These applications are designed to do something quickly, rather than conduct long communications sessions.

The format of a UDP segment is shown in Figure 2.13. As you can see, it's a much simpler structure than the TCP segment, but it does have some similarities.

Figure 2.13
The UDP segment structure.

Source Port	Destination Port
Length	Checksum
Data/User Payload	

As we noted earlier, both TCP and UDP use the concept of a *port* to associate data with an application, so port numbers are used. Both TCP and UDP generally use the same ports, where possible, for an application. UDP also performs simple error checking based on a checksum. It can provide some padding to ensure the data fits within 32-bit boundaries.

Within the last year or two, *instant messaging* (IM) applications have seen a tremendous growth in popularity. These applications use UDP for transport because they do not warrant the overhead required for a guaranteed delivery of information. We said that UDP relies on a higher-layer protocol or application to request retransmission if data is lost or corrupted. In the case of instant messaging, that request must come from the user. This is an example of using UDP for quick delivery of data. Although it isn't truly "instant," the protocol provides for rapid delivery where overhead is unnecessary.

It may seem like we've spent quite a bit of time dwelling on IP, TCP, and UDP when the subject of this book is IP telephony, but we've done so for a reason, and as we move into the chapters ahead, this foundation will be very important in broadening our appreciation of why some things have been done the way they have, and why VoIP may be a good idea in some instances with the technologies and advancements available today.

In the chapters ahead, we'll explore why voice packets might be carried in UDP, when it seems obvious that they should be carried in TCP for reliability. And we'll look at the use of TCP as a protocol to carry signaling information, much as the SS7 network is used in the PSTN.

Fundamentals of Packetized Voice

To transmit voice in IP packets it must be digitized. Earlier, we talked about the difference between analog and digital signals and some of the very basic transmission characteristics of each.

Voice digitization is not a new technology. The telephone companies began conversion to digital networks in the 1960s with the early deployments of T-1 service. At its inception, T-carrier, as it was called then, was devised as a trunking technology. Analog transmission technologies were widely deployed and in use at the time, and digital technologies allowed for some performance improvements that were beneficial to both customers and the phone companies.

In particular, during the 1960s, AT&T introduced *direct dial* long distance service. Prior to this service, a customer had to call a long distance operator to place a call outside the local calling area. The operator would then take whatever steps were necessary to provide the trunking connections to establish the circuit for the call. Digital networks offered a technology whereby customers could dial the call, and the digital signals that control the network could establish this connection "on demand," without the intervention of an operator.

Additionally, the time involved in setting up and tearing down circuit connections was causing performance issues in the network. Digital switches and technology provided a faster switching technology that was also cheaper, at a time when the economy in the United States was picking up and consumers were purchasing goods and services at a steadily increasing pace. Telephone service blossomed into more than just a method to communicate: it became an ingrained aspect of daily life and grew to be something essential and always available.

Before further exploring the technologies, we need to touch on standards, something that is a requirement for any technology that crosses more than one vendor or service provider.

A Word About the Standards

My friends and colleagues at Hill Associates often use the following story to describe standards. The actual source of this analogy remains something of an urban legend, and it is found cited to *Unknown* on many Internet resources.

In the United States the gauge, or distance between the rails on railroad tracks is 4 feet, 8.5 inches. That seems an unusual number to select. Why that particular gauge? It turns out that the same gauge is used in England and, given the way the United States was populated, English expatriates and immigrants made up the work force that built American railways to a large extent, But still, the question remains, why that size? Why did the English select that gauge? It turns out that the same people who built the pre-railroad tramways built the railroads, and that's the gauge they used for the tramways.

Again, why that particular gauge? Stepping back in history, the people who built the tramways used the same tools, patterns, and jigs that they used for building wagons, which used that distance as the wheel spacing. Again, the obvious question, why did the wagons use that spacing? England, and much of Europe is criss-crossed by old rutted roads. If the wheel spacing were different, the wagons would break on the old, long-distance roads, because that is the spacing of the old wheel ruts.

Why are the ruts spaced that way to begin with? Who built these roads? Roman legions built the first, and many of the long-distance roads throughout Europe. These roads have been used for centuries since. Built to the same width that was necessary for the Roman legions' war chariots. They were all built to identical wheel spacing for the Imperial Roman legions, and designed to be drawn behind two horses.

The answer to the question of railroad spacing can be directly attributed to the Roman legions. Specifications and bureaucracies take on a life of their own and live forever.

Although this story gives some insight into the long-term effects of standards, in simple terms today, standards provide interoperability. As technology has grown and flourished, standards have become the yardstick by which we measure the ability of a particular product or protocol to be used in a network that is supported by vendors, providers, countries, and companies around the world. Global interoperability depends on standards.

Several standards organizations are involved in the evolution and development of telecommunications standards, and we need to understand who they are and what role they play.

- The Institute for Electrical and Electronic Engineers (IEEE) provides a variety of standards related directly to technology. As the Internet evolves, this group provides LAN standards that are widely deployed for Ethernet, Token Ring, WiFi, and other net-related technologies.

- The American National Standards Institute (ANSI) is the U.S. representative to the International Organization for Standardization (ISO). ANSI really acts as an overseer of standards development and coordinates with the international ISO group. ANSI has done some development work on high-speed LAN standards. ANSI and ISO both develop, adopt, and support regional and national standards developed by other groups.

- The Telecommunication Standardization Sector of the International Telecommunications Union, or ITU-T was formerly known as the CCITT, or Comité Consultatif Internationale de Télégraphie et Téléphonie. This group is headquartered in Geneva, Switzerland, and is the "primary international body for fostering cooperative standards for telecommunications equipment and systems." The group operates under the auspices of the United Nations.

- The Electronic Industries Alliance/Telecommunications Industries Association (EIA-TIA) plays a crucial role in standards related to cabling, connectors, and hardware at the physical layer. Their predominant contribution has been in areas of building cabling and physical infrastructure.

- The Internet Society (ISOC) encompasses many different working groups. In the context of this book, the Internet Engineering Task Force (IETF) is the most relevant group. The IETF can be found on the Web at www.ietf.org and is responsible for organizing Internet-related standards. All Internet standards are open and published standards that are available online. The IETF manages the standards by a process of published Requests for Comments (RFCs), which are reviewed by a vast community of interest before being adopted as formal standards. Currently, over 3,000 RFCs, describing standards and proposed standards, are available from the IETF. The standard development process in the Internet environment is an important part of adopting new technologies like IP telephony.

A Telephone Call Simplified

When a user makes a telephone call, there are several basic steps that must occur. It doesn't matter whether the call is for a videoconference, to send a fax, or to carry on a voice conversation; these tasks have to be performed by the network in each case.

When we talk, the vibration from our vocal cords generates sounds waves, which are an analog signal. This sound is transmitted through the air, and our ears convert these sound waves back into signals our brain can understand (Figure 3.1). The world around us is very much an analog place.

The telephone is a device that converts the sound waves from our spoken conversation into electrical signals that can be transmitted over copper wires. The telephone set at the receiver converts that electrical signal back into analog sounds waves.

A Simple Telephone Call

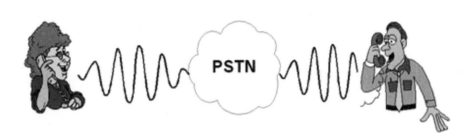

Figure 3.1
The network establishes the connection, encodes the signals, then transmits them and decodes the signal at the receiver. At the completion of the call, the network then tears down the connection and updates usage records to bill for the call.

In the telephone network, that electrical signal is a digital signal, a stream of 0s and 1s, defined by changes in the electrical state of the circuit. So, to carry on a telephone conversation, our analog voice must be changed from sound waves made up of vibrating air molecules into an electrical representation of the analog wave. This analog wave then must be digitized and transmitted across the network. At the other end, the signal is converted back to analog, then back into sound waves at the earpiece of the telephone handset.

Analog to Digital Conversion

Because the telephone network itself is made of digital central office switches connected by digital trunks, it makes sense to transmit digital signals. The local loops that connect subscribers and most telephone sets used on the PSTN are analog. Thus, somewhere an analog-to-digital conversion must take place. This conversion is typically performed using a technique known as pulse code modulation (PCM). To perform PCM, a coder and decoder are needed. These are combined into a single device called a *codec*. PCM samples a voice conversation 8,000 times per second. Each sample is converted into an 8-bit word, resulting in a 64 kbps sample (8 bits × 8,000 samples per second = 64 kbps). Each of these 8-bit samples can be coded into one of 255 different possible combinations. As in the binary math we reviewed in IP addressing, 8 bits of binary data can represent values from 0 to 255, but all zeroes cannot be used in this coding scheme.

We'll address the 64 kbps line rate more than once throughout this book, because it is the line speed provided by a standard voice channel in the PSTN. In the United States, a common transmission scheme using *time-division multiplexing* (TDM) is used to transmit 24 voice channels as a digital stream of data over a single circuit, which is the basic service provided over a T-1 circuit today—24 voice channels delivered over 1.544 Mpbs (due to some overhead). This design of digital facilities and voice circuits is part of the *Synchronous Digital Hierarchy* (SDH) used throughout the world. Other areas of the world do not use T-1 circuits, but the circuits used are very similar in function.

Sixty-four kbps really doesn't seem like a lot of bandwidth by today's standards, but the technical truth is that it's far more than necessary to carry a voice conversation. In an upcoming section, we'll look at some of the coding schemes used to sample and compress audio traffic.

The key factor in reducing the bandwidth required for a phone call really has to do with network economics. The smaller the bandwidth required per conversation, the more conversations the network can carry. Adding carrying capacity without comparable equipment additions can drive the cost of carrying a phone call down significantly.

Pulse Amplitude Modulation (PAM)

Voice communication is conducted via an analog signal, the human voice. The human voice is then converted into electrical signals. To carry this signal over the PSTN, we sample the amplitude of the signal at regular time intervals and create a collection of analog samples called a *pulse train*.

This PAM signal is still an analog signal, because it can represent any value. The technique just samples and encodes with no correlation to any discrete or fixed set of values. Nyquist's theorem on sampling demonstrates that we can sample at twice the highest frequency. Because the voice channel occupies 0 Hz to 4,000 Hz, we can collect 8,000 samples per second and accurately recreate the signal on the receiving end. Because the sample is an analog sample, it has the same constraints about transmission over distances as the original analog signal.

Figure 3.2
PAM signals—short duration amplitude pulses.

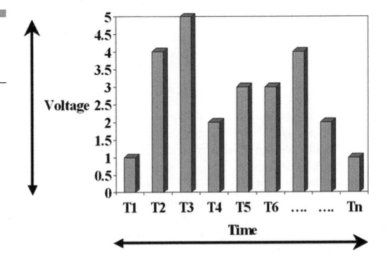

This analog PAM signal has to be digitized (Figure 3.2), which means that the infinite number of variable signals have to be rounded off into some discrete set of possible samples. If we were to use an eight-level approach, we could use 3 bits to encode each possible sample as a 4-bit digital value. This rounding off process is called *quantization*.

The rounding off process does add some error, called *quantization error* or *quantizing noise* (Figure 3.3) because we are rounding off each of the infinite values to some discrete number, but this error is generally undetectable to the human ear. The greater the *granularity*, or finer the scale used in quantization, the fewer errors and more accurate the representation. For voice traffic, 256 levels have proven adequate, so we use an 8-bit word to represent 256 possible samples in binary ($2^8 = 256$).

Another process, called *companding (compressing/expanding)* is used to reduce noise from quantizing. The human ear is more sensitive to changes in volume at lower levels than at high levels. Think of someone

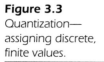

Figure 3.3
Quantization—
assigning discrete,
finite values.

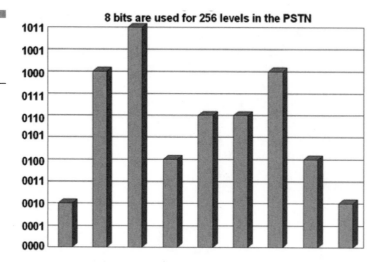

whispering in a quiet room. The slightest change in volume is quite noticeable. Now imaging sitting at a rock concert, trying to discern a slight change in the volume of the music. The human ear doesn't notice subtle changes at higher levels.

We take advantage of this human physiology by focusing engineering efforts on reducing quantizing noise or error to those portions the ear is more likely to discern. In other words, we sample more frequently at lower amplitudes (or volume) and less frequently at higher ones. *Companding* is the process of uneven encoding, using closely placed samples at low amplitudes and widely spaced samples at high ones.

In the early 1970s, the CCITT, which has now become the ITU-T, worked on standardizing voice processing in the global PSTN. *G.711* is the ITU-T standard for digital speech encoding of voice, as described in this section. The process of negotiating and establishing international standards is always fraught with technological and political disagreement, and a single, universal standard for companding could not be achieved. Japan and North America wanted to use a standard referred to as *Mu-law* (μ-255), whereas Europe favored *A-law*, another 8-bit approach, but one that inverts the even bits. While both use 8 bits, the bit string is coded differently. As a result, both schemes are part of the international G.711 standards for PCM voice today.

Because international traffic must now use A-law companding, this requires conversion back and forth in the PSTN to support the technology in use in a particular country.

The Complete Digitizing Process

The complete process, referred to as *pulse code modulation* (PCM), is shown in Figure 3.4. It consists of four steps to ready the signal for transmission.

1. *Filtering* unwanted frequencies is the first step. Only those frequencies in the 0 Hz to 4,000 Hz range need be sampled because that's the range used to transmit voice. This is also sometimes referred to as a *4 kHz voice channel*.
2. *Sample the analog signal* using Nyquist's sampling theorem. Because the maximum bandwidth is 4,000 Hz, we sample 8,000 times per second. The output of this is a series of analog pulses called the PAM signal.
3. *Quantizing using a companding scheme* (μ-255 in the United States, A-law in Europe) to assign a discrete digital value to each sample.
4. *Pulse code modulation (PCM)* is the process of assigning an 8-bit value (called a PCM word).

Sampling at a rate of 8,000 samples per second and encoding each sample into an 8-bit word results in a bit rate of 64 kbps for each telephone conversation. This DS-0 signal level is commonly used throughout the PSTN worldwide.

Figure 3.4
Digitizing voice
signals.

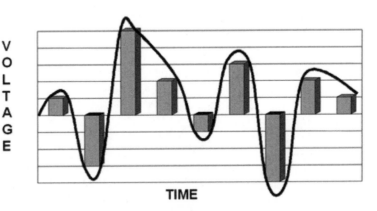

Filter unwanted frequencies. Only sample 0 – 4 kHz
PAM Samping, 8000 samples pers second
Quantizing 256 amplitude levels into discrete companded signals
Pulse Code Modulation (PCM) encodes the samples into 8-bit words
8 X 8000 = 64 kbps

Although understanding these coding schemes may not be viewed as crucial to the process of packetizing voice traffic for IP, it is important, particularly because most IP telephony solutions today use and support PCM encoding as the method for digitizing voice traffic. Additionally, it points to the reasoning behind the 64 kbps bandwidth limitation so prevalent in the telephone network today. We must remember that the PSTN was designed and tuned for optimal transmission of this voice signal. As a result, some of the engineering techniques used to provide for better voice delivery mechanisms have now proved problematic in the delivery of data transmissions.

IP Telephony versus Traditional Telephony

Although the technologies used are similar in many aspects, fundamental differences exist between IP telephony and traditional telephony. In Figure 3.5, the bottom portion shows a simple traditional telephone call over the PSTN. The subscriber picks up the telephone and receives a dial tone from the local phone company CO switch. Calls might be made within the local phone company serving area or connected long distance via an *interexchange carrier* (IEX or IXC) to another local company serving another area. In that event, trunks between the COs are connected to establish the path for the call. Regardless of where the call is destined, these PSTN telephone calls are *circuit-switched* connections, establishing a dedicated path for the duration of the call. Nobody except the calling and called parties may use this path or these resources in the network until the call is completed. Once the call is completed, all connections are released and the network databases are updated to charge appropriate billing.

Within the PSTN, Signaling System 7 (SS7) performs the necessary programming translations to establish the circuit path and log billing information. In reality, several circuits are connected together end to end to complete a connection. The local phone company, or *local exchange carrier* (LEC) performs analog to digital conversion, with the interexchange portion remaining a digital connection. The IEC provides long haul transport of the call over greater distances between calling areas. The IEC pays access charges to the LEC for termination of long distance calls in the LEC area.

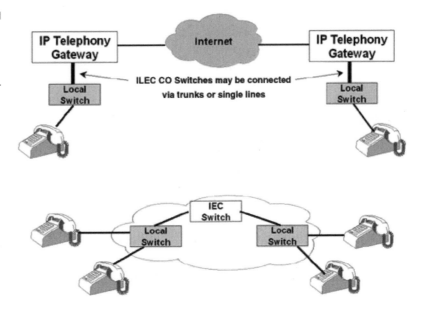

Figure 3.5
IP telephony
compared to
traditional telephony.

The upper portion of Figure 3.5 demonstrates an IP telephony call using the Internet as a transport network. In this environment, the calling and called parties do not directly use IP. Their telephones are connected to a traditional CO switch, which provides all the standard calling features. When a call is placed from the switch on the left to the one on the right, the CO switch routes the call to an *IP telephony gateway*. A *gateway* is a device that converts information from one kind of network to another. In this case, the gateway converts standard PSTN signaling and voice traffic into IP packets, which are routed across the Internet to the destination gateway on the far right. That gateway converts the IP packets back into normal PSTN signaling and digitized voice traffic, which gets sent to the local switch. In turn, the caller's phone rings and the call is completed.

The significant, and it is very significant, difference is that the "cloud" portion, the Internet, is a packet-switched network. No paths are established, and there are no dedicated resources. Everything in the network is shared among all users.

Consider an average telephone call and how much *dead time*, silence, or space between words consumes time over the dedicated PSTN connection. In the packetized world, perhaps we need not transmit silence, thus freeing up the bandwidth to carry packets sent between other users. This allows for the potential to recoup tremendous amounts of bandwidth in the network.

Gateways are very complex and sophisticated devices, with a wide range of features and capabilities available; we explore them further in another chapter.

Voice Quality in the Network

Even though the PSTN is over 100 years old, voice quality is still measured by placing a group of people in a room and having them listen to sound in headphones. The evaluators rate the quality of the sound from 1 to 5. A 5 is the highest rating, being what might be called "pin-drop" quality. Many providers refer to this as "toll-quality voice." This highest-grade voice quality has always been the benchmark for conducting corporate business calls. A rating of 1 equates more to the scratchy sound quality of an intercom speaker in a warehouse or at the drive-through hamburger stand. This rating, from 1 to 5, is referred to as the *mean opinion score* (MOS). Although the statistical validity may be questionable to some, the process has worked satisfactorily for a number of years, and this method is accepted worldwide. In the real world, the human ear can clearly distinguish between a 4 MOS and a 4.5 MOS.

An MOS of 4 to 5 is considered toll-quality voice and rated suitable for the long distance business world to use in negotiating business deals. Ratings below a 3 are generally considered to be synthetic in quality, and may be referred to as a "robotic" sounding voice. This has always been important because the fewer the number of bits used in the encoding scheme, the poorer the quality of the voice. Because the PCM encoding scheme drives the network to a 64 kbps voice channel, a sampling algorithm that requires less bandwidth could result in improved network efficiency. With the enhancements in digital signal processor (DSP) technology and improvements in electronics, newer sampling techniques are becoming cheaper and easier to implement.

Figure 3.6 represents a comparison of several encoding schemes that are in common use for various applications. It lists the codec types and algorithms used, the bit rate and sample size, the algorithmic encoding delay, and then compares the mean opinion scores for various digitization approaches.

The *encoding delay time* is applicable to the algorithms used. Many factors influence delay, but the processing time of the algorithm itself must be considered in terms of the total system delay. In any IP network, the overall transport delays of moving packets are unpredictable

Figure 3.6

Encoding scheme comparisons.

ITU-T Codec Standard	Coding Scheme Used	Bit Rate	Sample Size (Bits)	Encoding Delay Time	Mean Opinion Score
G.711	PCM	64 kbps	8	<1 msec	4.4
G.722	SB-ADPCM	64 kbps	8	4 msec	
G.726	ADPCM	32 kbps	4	1 msec	4.2
G.728	LC-CELP	16 kbps	40	2 msec	4.2
G.729	CS-ACELP	8 kbps	80	15 msec	4.2
G.723.1	MPMLQ	6.3 kbps	192	37.5 msec	3.98
G.723.1	ACELP	5.3 kbps	160	37.5 msec	3.5

and variable. These factors alone may make the network unsuitable for real-time voice traffic. The nodal processing delay involved in encoding and decoding could add enough overhead to the end-to-end delay that the threshold of acceptable service is crossed and the network becomes unusable. The sum total of all the delays cannot exceed 300 milliseconds for an interactive voice network, and many providers strive for 200 milliseconds total delay or less.

Although the algorithms are listed and described, we will not undertake explaining how each works in detail. Each algorithm is documented in ITU-T standards and a wide variety of papers. Interested readers should search the Internet for more details on the specific mathematical workings of the algorithms.

Pulse code modulation, or G.711, is the approach used in the PSTN today, and while it's advertised as being toll or "pin-drop" quality, it still falls somewhat short of the perfect score of 5. The 64 kbps PCM time slot provides the basic framework for contemporary public telephone services and equipment. This encoding scheme was widely used in most early IP telephony systems, and is supported by virtually every equipment vendor in the VoIP sector.

The G.722 codec is used for FM radio and does not have an MOS associated with it. It is included for comparison purposes as a means of encoding sound waves for transmission.

Adaptive differential pulse code modulation (ADPCM) offers a solution that could reduce the bandwidth requirements by half while only sacrificing .2 on perceived quality in the MOS.

Low-delay code excited linear predicate (LC-CELP) coding has been widely used in voice mail systems for digitizing stored voice messages.

David Isenberg of Bell Labs did much to facilitate this in Release 2 of the AUDIX voice mail system at AT&T in the late 1980s.

Conjugated structure algebraic code excited linear predictive (CS-ACELP) can deliver an 8-kilobit sample with less than 16 msec of processing time. This G.729 codec standard has been widely used in digital telephony, satellite transmission, and wireless communications. It is also used in Voice over Frame Relay (VoFR) and supported by many Frame Relay equipment vendors.

Multipulse maximum likelihood quantization (MPMLQ), while published as an ITU-T standard, has also seen several proprietary implementations with smaller samples, as low as 4.8 kbps. MPMLQ is able to maintain reliable performance despite a high bit error rate and has been deployed in many Russian telephony implementations over data networks.

Algebraic code excited linear predictive (ACELP) coding can produce a sample that has a bit rate of only 5.3 kbps. This approach has been deployed in many Frame Relay networks. It can be adjusted to encode at several bit rates. At 8 kbps, ACELP measures a 4.2 MOS, and it's able to adapt its rate on the fly. This capability could provide a mechanism for adapting to a network that doesn't offer consistent, predictable performance. ACELP actually creates models of the human voice, then predicts what the next sound will be. It encodes the difference between the actual sound and the predicted sound, and the difference is transmitted to the receiving end. Since the other end of the call is also running ACELP, the calculation of the differences allows for an acceptable recreation of the human voice at the receiving end. In the past, there were some complaints that ACELP techniques created a less accurate representation of women's and children's voices, which are generally higher in pitch than male voices. With improvements in digital signal processors, this drawback is less of an issue today.

Although PCM is still the most widely used voice digitization technology, the importance of these encoding schemes cannot be overemphasized. PCM creates an 8-kilobit sample, which requires 64 kbps from the network for transmission. This is the fundamental cornerstone of the T-carrier architecture in the Internet today. A T-1 channelized for voice traffic can carry 24 simultaneous telephone conversations; however, a T-1 data circuit carrying IP packets compressed at 8 kbps could arguably carry 192 conversations in the same amount of bandwidth.

When Does Packetized Voice Become Telephony?

As we've seen, the PSTN is a complex and mature implementation of technology that has evolved over more than 100 years of use. The Internet is still immature in years by comparison.

The advanced intelligent network (AIN) is part of the PSTN. An 800 number service provides access to databases of caller information. Caller ID, Call Waiting, Call Hold, and Conference Calling are common features available to almost all users. E911 services are changing to identify the caller's location to within a very small geographic radius. This mature network has been designed over time to provide a host of services that we now take for granted. The PSTN does far more than transport voice traffic. It provides a comprehensive and robust suite of telephony services.

Compare that with the IP network as we know it today. IP certainly can't provide a similar set of services by itself. VoIP has been achievable for a number of years, but there is far more to IP telephony than just carrying voice traffic in IP packets: a full telephony feature set has to be present, at least to an acceptable level, before the market begins to accept IP telephony as a viable product for deployment.

IP telephony bridges the differences between these two networks, using the TCP/IP protocol suite to transmit voice messages over a network like the Internet instead of the PSTN. The IP telephony market has seen a substantial growth rate over the past two years, and products and networks have improved. Predictions for the IP telephony market estimate revenues in excess of $100 billion.

This growth in IP telephony brings implications for everyone who has any involvement in the telephone network, from consumer to equipment vendor.

Early prophecies saw the success of IP telephony as a tool for consumers to combat the high price of long distance telephone service. While the arguments seemed solid at the time, they failed to take into account either the technical advances or the growth of the wireless market. Today, many wireless providers offer nationwide calling at rates considerably lower than were available when IP telephony first gained attention. However, significant interest still exists in international long distance using IP telephony as a direct result of international pricing and regulations.

In some cases, incumbent local exchange carriers (ILECs) and interexchange carriers (IECs) have experienced market erosion due to new, next-generation telco services, but most have begun their own explorations into integrating IP into their traditional voice network in ways that make the best effective use of the technology and result in the lowest cost for delivery of services. This all adds up to reduced cost to consumers.

Regulatory differences between the telephone industry and the Internet industry have played a crucial role in the development of IP telephony services. Telephone companies are very heavily regulated by the federal government, state government, and in some areas, counties and municipalities. The Internet remains a predominantly unregulated marketplace, with data networks treated as an unregulated information service.

As new alternate distribution companies spring up to offer middle-mile services, the door is open to new service providers offering a variety of last-mile services that includes both voice and data. This new competition, particularly among providers using wireless or cable technologies that don't require the telco local loop, could present a serious danger to the ILECs, whose business is dependent on the delivery of local telephone services.

Interexchange carriers are often also Internet providers and carriers of backbone traffic. Consolidation of multiple traffic types onto a single backbone infrastructure can result in tremendous cost savings and increased efficiency for these companies. Not only are there savings in equipment, but the support personnel required for a consolidated infrastructure can also be drawn from a single skill set rather than having specialists in both voice and data backbone technologies.

Equipment manufacturers have broadened their focus as well, with traditional telephony vendors like Nortel Networks and Lucent Technologies expanding into broader Internet equipment with the purchase of data networking companies. Cisco Systems has notably expanded in the other direction, spreading from a rich data background into voice technologies.

New vendors have discovered a technology sector that is wide open for competition, and relative newcomers such as Voyant Technologies and Shoreline Communications have introduced a variety of new hardware and software solutions. With the advances in networking technology, vendors now provide telecommunications technology on an array of platforms ranging from traditional PBXs to server-based telephone systems.

Overall, this has opened the way for greater competition in premise technologies, local technologies, and backbone technologies. Coupled with high-speed data services delivered over DSL and cable modem, we now see IP telephony-based distributed call centers being deployed as a next-generation telecommuter working environment. The opportunities for competition among the telephony service providers, and new uses of the offerings in business, have created an environment of innovation and competition that can only benefit end users.

IP Telephony Protocols

Earlier, we reviewed the need for standards and why they exist. We've discussed several different standards organizations. Now we're going to explore some of the standards being used and developed specifically for IP telephony. To keep things in perspective, we must first set the stage.

In the previous chapter, we mentioned that standards used in the worldwide PSTN are the work of the International Telecommunications Union—Telephony sector (ITU-T), formerly known as the CCITT. This group operates under the auspices of the United Nations, and that is an important note. This is a global body, and while it has responsibility for telephony standards, it has not historically been known for speed. Many different international political agendas come into play when dealing with the United Nations, and change takes time.

Standards in the Internet are primarily the work of the Internet Engineering Task Force (IETF). This is also a global group; however, it's primarily an organization of volunteers. Any interested party can join the IETF and participate in standards development. The IETF working groups are made up of technology specialists from colleges and universities, vendors, telecommunications providers, Internet providers, governments, and a variety of interested parties. The organization is voluntary, and the structure of developing standards is much different from that used in the PSTN.

Internet standards are developed through the Request for Comments (RFC) process, which is quite efficient. In many cases, interested parties band together to jointly present new open standards for the improvement of the network. It's clear that a proposal for a new standard jointly presented by a team of vendors and providers can carry broad support at introduction. Therefore, in the real world of standards development, change can often occur very quickly in Internet standards.

Thus, we have are two different networks, with technical standards developed and approved by two separate organizations. And while that sounds straightforward, it isn't. We often draw the network as a cloud, for the sake of simplicity, because lay people often don't need to know or understand the inner workings of the public network. Although this is true, even a concept as simple as a cloud raises complex problems where interoperability is a concern, and interoperability with the incumbent network, the PSTN, is absolutely necessary for global success and widespread deployment of IP telephony solutions.

In Figure 4.1 we see three distinct viewpoints. At the top, the PSTN and Internet are distinctly different networks, with no connection or interoperability required between the two. They perform different functions and are designed for different purposes. This model represents the

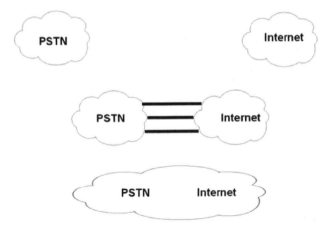

Figure 4.1
Three variations of
PSTN and Internet
networks.

relationship of these two networks in the 1970s and much of the 1980s. They operated independently of one another for the most part, and many felt that the two networks would never meet.

In the center, we see the two networks joined, or connected, not by three connections, but by thousands of links. When consumers started using modems to dial up to the Internet for access, the two networks interleaved to some degree. Later, as IP telephony began to blossom as an idea, it became clear that an Internet "phone call" would have to be compatible with the PSTN in order to provide ubiquitous service. Connectivity is not enough to allow for telephony; interoperability is a paramount issue. This conclusion presented the IETF with a mandate to work in cooperation with the ITU-T in ensuring that evolving Internet standards support and remain compatible with PSTN standards, and that the two networks function together, ideally in a seamless fashion. The phrase "transparent to the end user" has often been used to describe this interworking functionality. This perspective of cooperatively working together was the dominant viewpoint of many developers of the early IP telephony standards.

During the past two years, the word *convergence* has taken on a life all its own. At a high level, we think of convergence as the migration of services like voice, data, and video to a single consolidated network like the Internet. Some people said the Internet was growing so quickly it would absorb the PSTN, and others speculated that telephony would migrate to the Internet and that someday the PSTN would just be "turned off"—a rather bold forecast.

The real truth is that both networks are vital, growing, and critical. At one point, the common view was that the two would become linked together at high-capacity access points, passing traffic through gateways to each other. Today, they are far more inextricably interwoven than that. The two networks aren't one today, but they are so tightly interwoven into delivering the services required by business that they are beginning to represent a single cloud. This cloud is nothing more than service capacity, with services being defined through agreements between users and providers. The network of tomorrow—perhaps today—is a cloud of capacity that provides whatever service the end user requests of it. (For a more detailed review of this concept and the evolution of the networked world see Steven Shepard's book *Telecommunications Convergence*.)

As we've already mentioned, Internet standards, as developed by the IETF, are many and varied. In this chapter, we explore a few of the most commonly used and most vital standards dealing directly with the delivery of voice or multimedia services for IP telephony.

H.323 Standards for Multimedia Over Packet Networks

In 1995, the ITU-T began work on a series of standardized signaling protocols. One outgrowth of this work was a product called the *Internet Phone* from VocalTec. Initially, this was a proprietary solution and did not focus on interworking between the PSTN and the Internet. Transmission of audio signals, and the idea of videoconferencing over the Internet, were key issues. These standards fell within the H.323 family of protocols for multimedia transmission over packet networks, and were to some extent an outgrowth and extension of H.320 ISDN videoconferencing standards.

Interworking with the PSTN quickly became a major focal point of this technological development work. Designers recognized the need to incorporate some method for calls to traverse from the IP network to the PSTN. This led to efforts in gatewaying protocols and connectivity to the SS7 network, which provides extensive command and control functionality in the PSTN. *H.323* products began to appear from vendors in 1996.

H.323 embraces a set of goals that are quite simple and straightforward in principle; however, implementation has proven far more complex.

- Internetworking with the PSTN became a central theme.
- H.323 had to handle the conversion of signaling from whatever packet protocols were used to the PSTN signaling format used by SS7.
- H.323 had to have a call control mechanism for call setup and teardown.
- H.323 had to encode the media—digitize and packetize the audio voice transmission—for the IP network.

The last three functions were ideally performed in a single device referred to as a gateway, but overall H.323 encompasses a complex suite of protocols and approaches to signaling, media conversion, registration, and admission to the network as shown in Figure 4.2.

Figure 4.2
The scope of H.323.

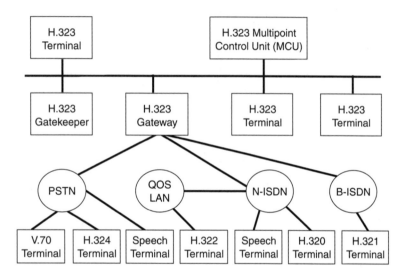

A gateway may support one or more of the PSTN, N-ISDN, and/or B-ISDN connections.

H.323 is often referred to as an "umbrella standard" because it includes standards for a variety of purposes. Codecs are defined for the transport of both audio and video signals. Audio codecs are included for compression as low as 5.3 kbps for voice. It is important to note that "voice" can mean audio, modem data, fax messages, or touch tone signals (DTMF). Encoding schemes don't necessarily work equally well with each of these. The *Real Time Transport Protocol* (RTP) is defined for use with both audio and video. It is used for delay-sensitive information and includes timestamping for the sequencing and timing of packet delivery. Whenever RTP is used, *Real Time Control Protocol* (RTCP) is also used. This control protocol establishes and monitors RTP sessions.

It is important to note that RTTP uses UDP for the transmission of the audio and video packets. Because this medium is providing for delivery of real-time information, UDP is used for the quickest delivery. TCP can provide guaranteed delivery of information, but the overhead associated with TCP, coupled with the retransmission of any lost data, is too intrusive to support real-time data delivery. Therefore, a voice packet that is lost during transmission is simply lost. Testing has shown that in real-world applications, the human ear is far more tolerant of lost packets containing a fraction of a syllable that of the delays of using TCP.

The *H.225 standard* is used for *registration, admission,* and *status,* or RAS. An H.323-compliant terminal, upon connection with the network, registers with a *gatekeeper* in order to participate as a member of the voice network. Stations need to request network resources, make calls, and resolve the IP address of the called station. A gatekeeper often performs these functions, although the device is optional under the H.323 standards.

Q.931 signaling is used between devices for call setup and teardown. Q.931 is the identical signaling protocol used in ISDN services, complete with all the features used there. This signaling can be supported by a gateway, but can also be supported by a telco central office (CO) switch. This signaling has provided a crucial interoperability function in allowing signals to move between IP networks and the PSTN.

H.245 standards are used to provide an exchange of media capabilities between end stations. The H.323 family of protocols supports multimedia, including video. H.245 might be used when a video call is requested to negotiate for voice-only connectivity, if the called party does not have a video-capable terminal.

According to the standards, *reliable transport* is used for signaling. Thus, in an IP network, the overhead and performance of TCP are necessary. A good way to view this is to correlate this signaling to the support provided by the SS7 network in the PSTN. Without guaranteed delivery of call control and signaling message, call processing would halt and the system could not function.

The *T.120 standard* was implemented in 1996–1997 and contains protocols and services that support real-time, multipoint data communications. T.120 has often been implemented in the form of a "whiteboard" application that both users can share, but it is also used for sharing files or multiplayer gaming. One example might be two users collaborating on a spreadsheet while talking about the changes being made. Many vendors, including Apple, AT&T, Cisco Systems, Intel, MCI, and Microsoft have implemented and support T.120-based products or services.

Figure 4.3 demonstrates the relationships between these protocols.

Figure 4.3
H.323 protocols.

Video	Audio	Control			Data
H.261 H.263 (video codec)	G.711 G.722 G.723 G.728 G.729 (audio codec)	H.225 Terminal to gatekeeper signaling	Q.931 Call signaling	H.245 Control Channel	T.120 (Data terminal sharing)
RTP	R T C P	RTP	R T C P		

Call Setup Using H.323

Many people have suggested that H.323 is cumbersome. Products supporting it have often been accused of "software bloat" because of the comprehensive functionality included and supported. Yet H.323 is widely deployed and supported today. Given that, we'll explore what it takes to establish a telephone call between two workstations using H.323 protocols.

The scenario we review is rather simple. Two users in Figure 4.4, Bob and Alice, in the same network, on the same LAN, in the same building, wish to communicate. Bob needs to call Alice to discuss a project they are working on. We assume that there is a gatekeeper on the network for administration purposes, and step through the entire process of establishing the necessary connections and sessions to conduct a telephone call. We assume that they are both using their computers as the H.323 workstation or telephone:

1. When Bob and Alice boot up their computers in the morning, each station must send a Discovery message to locate the Gatekeeper.
2. The Gatekeeper must reply and provide its IP address

3. Each workstation must now transmit a Registration Request to the Gatekeeper and receive an Acknowledgment in return.
4. Since Bob wants to call Alice, his workstation sends a Locate Request and receives an Acknowledgment in return. This provides the IP address of Alice's workstation.
5. Bob's station now transmits an Access Request seeking resources and permission to make the call and receives an Acknowledgment in return.
6. At this point, a TCP session is established for H.225 setup.
7. Alice's workstation has received an incoming call request and now must transmit an Access Request for resource and receive Acknowledgment.
8. An H.245 Connect message is exchanged between the two workstations.
9. A second TCP session is established for the H.245 session.
10. Bob's workstation must open a Logical Channel for the media stream in the forward direction and receive an Acknowledgment.
11. An RTP media forward channel is opened, immediately followed by an RTCP control stream in the reverse direction.
12. Since the stations negotiated a full-duplex, or two-way call, at the capabilities exchange phase, Alice's workstation must open a Logical Channel for the media stream in the reverse direction and receive an Acknowledgment.
13. An RTP media reverse channel is opened, immediately followed by an RTCP control stream, in the forward direction.
14. We have established a full-duplex, two-way path for voice audio.

If this seems complex, keep in mind that H.323 supports multimedia beyond pure voice. If full-duplex video were negotiated, steps 10 through 14 would be necessary to set up those media streams. If data sharing were also required, steps 10 through 14 would have to be repeated again to establish that stream, and another TCP session would be established for T.120 data.

As you can see, establishing a call in H.323 can be overhead intensive. Although this works very quickly on a local LAN segment, where bandwidth is not a real problem, over a wide area network like the Internet, the delay in call setup can pose a serious problem. In some cases, setting up an H.323 call can take several seconds. Because TCP has timers for the retransmission of packets, the delays can be much worse if the network is suffering from congestion and packet loss.

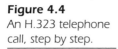

Figure 4.4
An H.323 telephone
call, step by step.

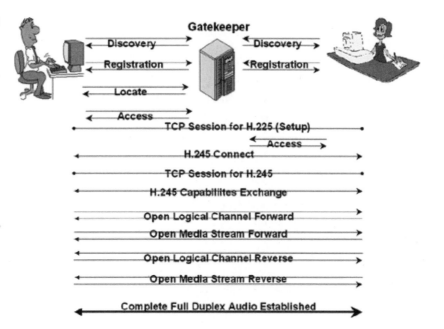

H.323 is widely used and supported by every major equipment vendor. This support remains to maintain full compatibility with the PSTN and ITU-T standards, but designers quickly observed that the complete feature set of H.323 might not be necessary for IP telephony. And the inclusion of the complete set of protocols led to products that were large in size, and often inefficient in their coding and use of system resources.

H.323 Version 2 allows for fast connections, and supports opening media streams simultaneously, but the issue of overhead remains a concern to many designers of voice services and software.

Note that none of the discussion so far has provided a mechanism for call transfer or diverting, call hold, call parking, call pickup, call waiting, or message waiting services, all common services in business systems and expected to be present in today's working environment. Neither is there a mechanism for failover to an alternate path if a node becomes congested or does not have resources necessary to support the call.

Many designers, the author included, argue that whereas H.323 provides the necessary functionality, there are far too many scalability and performance-related issues to treat it as an acceptable solution for current IP telephony technology. It is included here because of its incumbent position in many applications and implementations, but clearly better performance is necessary for widespread success of IP telephony.

Session Initiated Protocol (SIP)

Unlike H.323, the original work on Session Initiated Protocol (SIP) was performed by the IETF as one of several different efforts. The Multiparty Multimedia Session Control (MMUSIC) working group took much of the lead in early efforts. Since 1999, the IETF-SIP working group has led this work. Their specific charter states "SIP is a text-based protocol, similar to HTTP and SMTP, for initiating interactive communication sessions between users. Such sessions include voice, video, chat, interactive games, and virtual reality. The main work of the group involves bringing SIP from proposed to draft standard, in addition to specifying and developing proposed extensions that arise out of strong requirements. The SIP working groups concentrate on the specification of SIP and its extensions, and will not explore the use of SIP for specific environments or applications.

Throughout its work, the group strives to maintain the basic model and architecture defined by SIP. In particular:

1. Services and features are provided end-to-end whenever possible.
2. Extensions and new features must be generally applicable, and not applicable only to a specific set of session types.
3. Simplicity is key.
4. Reuse of existing IP protocols and architectures, and integrating with other IP applications, is crucial."

SIP provides protocols and mechanisms that allow both end systems and proxy servers to provide services including:

- Call forwarding under a variety of scenarios (no answer, busy, etc.)
- Calling party and called party number identification using any naming scheme
- Personal mobility allowing a single address that is location and terminal independent
- Capabilities negotiation between terminals
- Call transfer
- Instant messaging
- Event notification
- Control of networked devices

Extensions to SIP also provide for fully meshed conferences and connections to multipoint control units (MCUs).

SIP uses an addressing structure very much like email addressing. Given that a user might log on from any location and receive an IP address dynamically, there must be a way to resolve common naming conventions to the active and current IP address. Because people are familiar and comfortable with email addresses, this structure seems most appropriate and remains a popular choice.

Because SIP is a text-based protocol like HTTP or SMTP, the addresses, which are SIP uniform resource locaters (URLs), can be embedded in email messages or Web pages. Also, since this is a text protocol, the addresses are network-neutral, thus the URL might point to an email-like address, using SIP, an H.323 address, or it might point to a PSTN telephone number. The ITU-T E.164 standard defines the telephone numbering structure.

SIP provides a comprehensive set of building blocks that can be extended to allow for E911 or advanced intelligent network services.

Because it can support forking to multiple destinations, SIP can support call forwarding, automatic call distribution (ACD) techniques for call centers, and redirecting the call to multiple alternate locations.

SIP operates independently of the Network Layer and requires only unreliable datagram or packet delivery. It provides its own reliability mechanism. While in the IP environment of the Internet SIP is used over UDP or TCP, it could run over IPX, Frame Relay, ATM AAL5, or X.25 without modification. Generally UDP is used to avoid the overhead associated with TCP.

The protocol model for SIP, as described in RFC 2543, is shown in Figure 4.5. It provides four different functional components:

- *User agents* (UAs) either initiate call requests or are the destination of those requests. A user agent might be IP telephony software running in a computer or an IP telephone.
- The *registrar* keeps track of users within the network or domain. User agents register with the registrar as members of the network.
- The *proxy server* is an Application Layer routing process that directs SIP requests and replies within the network.
- The *redirect server* receives requests for users (UAs) and provides the location of other SIP user agents or servers where the called party can be reached.

Within the SIP server, the registrar, proxy server, and redirect server can be implemented in the same software package.

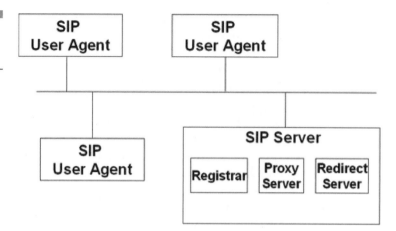

Figure 4.5
Session initiation protocol model.

During a SIP session, a user initiates a call, which prompts the user agent to transmit a SIP message. These messages traverse one or more SIP servers. Once the destination user agent information is obtained, actual message transfer takes place directly between the user agents. If one end of the call is located in the PSTN, a gateway between the IP-based SIP network and the PSTN is required to provide all the necessary protocol conversions between networks.

Session Description Protocol (SDP)

The MMUSIC working group of the IETF also provided RFC 2327, the Session Description Protocol (SDP). SDP is intended to describe multimedia sessions for the purposes of session announcement, session invitation, and other forms of multimedia session initiation. The Session Description Protocol is used within both SIP and Megaco implementations. SDP is not intended to support negotiation of session content or media encoding, but to act as a general-purpose tool. It also supports multicast media and can be used for broadcast environments like Internet radio or television.

SDP provides *session announcements* as the mechanism used to convey session description information between devices or nodes and proactively deliver these to users. These announcements might also be delivered via email or the Web, allowing for automatic launching of the appropriate application on the called party's workstation. SDP includes:

- The sessions name and purpose
- The time the session is active
- The type of media used in the session; this might be voice, video, data, etc.
- The format of the media (MPEG video, H.261 video, etc.)
- The transport protocol used (UDP, TCP, IP, etc.)
- Information necessary to receive the media (TCP/IP ports, addresses, and formats)

The actual syntax for the port and addressing information varies depending on the transport protocol in use. The following is an example of an SDP description:

```
v=0
o=kcamp 2890844526 2890842807 IN IP4 126.16.64.4
s=SDP Seminar
i=A Seminar on IP Telephony
u=http://www.ipadventures.com/seminar/voip.17.ps
e=ken@ipadventures.com (Ken Camp
c=IN IP4 63.215.128.129/127
t=7944393265 8746931596
a=recvonly
m=audio 49360 RTP/AVP 0
m=video 51782 RTP/AVP 31
m=application 32416 udp wb
a=orient:portrait
```

In general terms, SDP is used to convey enough information for a user to join the session or call. This may not include encryption keys in the virtual private network (VPN) environment, which might be handled by another set of protocols like IPSec.

We won't explore the inner workings of SDP further, but it's necessary to have a high-level grasp of how the information describing multimedia sessions is transmitted between user systems.

Call Setup Using SIP

Earlier, we stepped through a telephone call between Bob and Alice. Now we'll step through that process again, using SIP as our IP telephony protocol. We'll assume each user is on a different LAN and the two networks are connected via a router. Each network has a SIP server on the local LAN, segment in this scenario.

1. At startup of the user agent software (the IP telephony package), both Bob and Alice automatically register with their local SIP server.
2. Bob initiates a telephone call. In so doing, the user agent on Bob's computer transmits an "invitation" to the SIP server on his local network. This invitation contains the session description information. In most cases, rather than using some discovery method, SIP servers are statically configured in the SIP user agent software.
3. Since Alice registered with the SIP server on another network, the SIP server on Bob's network doesn't know how to reach her. It forwards the invitation to every SIP server it knows how to reach—in this example, Alice's SIP server.
4. Since Alice is on the same LAN and registered with her SIP server, it knows how to reach her and forwards the invitation to her.
5. Since Alice also wants to talk to Bob, she answers that call, which returns and acknowledgment (ACK) back over the same path the invitation followed. Alice's session description information is included in the acknowledgment.
6. Since both ends have exchanged session description information, they have the IP address and port information to directly contact the other party on the call. They can now transmit RTP encapsulated media directly. The SIP servers do not need to participate in the call session any further.

Figure 4.6
An SIP telephone call, step by step.

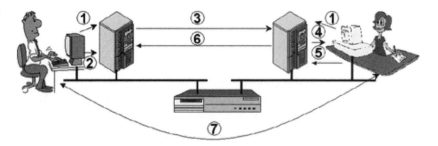

① Registration – performed by each station

② Bob initiates call, sending invitation and SDP

③ SIP Server forwards invitation to all known SIP Servers

④ Alice's SIP Server delivers invitation to called party

⑤ Alice accepts invitation and returns SDP

⑥ SIP Server returns ACK to calling party

⑦ End-to-end telephone conversation

As you can see in Figure 4.6, the process is simpler than H.323, using fewer messages for call setup. It doesn't require TCP, which can eliminate more overhead for improved performance. Beyond that, many designers feel that the registration process with SIP servers provides better support for mobile users. And because SIP is a text-oriented protocol, a simple BYE command is used to terminate the session.

Comparison of H.323 and SIP

Table 4.1 compares some of the features and functionality of H.323 and SIP side-by-side. This is not a complete or exhaustive comparison, but gives us a quick look at some key similarities and differences between the two.

TABLE 4.1

H.323 versus SIP

	H.323	SIP
Transport Protocol(s)	Uses both TCP and UDP. Requires reliable transport.	Can use either TCP or UDP. Can also run on any unreliable packet network.
Addressing format	Allows addresses to hosts directly. Aliases resolved by a gatekeeper.	Address-neutral URL, including email, phone, H.323, and HTTP. Email-like names can be mapped to any network device.
Multicast	Supported by another H.332 set of specifications.	Caller can invite called party to join multicast sessions.
Topology	Uses gatekeeper routing and has no loop detection.	Supports fully meshed, multicast, and MCU-based conference calling, including loop detection capability.
Complexity	More complex call setup.	Simplified call setup.
Mobility support	Limited support for mobility.	Supports call redirection, call transfer, and similar telephony features.
Authentication	No user authentication is included.	User authentication can be performed via HTTP, S-HTTP, SSH, or any HTTP-like transport layer security.

continued on next page

TABLE 4.1

H.323 vs. SIP

	H.323	SIP
Protocol Encoding	H.323 uses Q.931 and the ASN.1 PER encoding.	SIP is a text-based protocol similar to HTTP.
Connection state	H.323 calls can be terminated explicitly or when the H.245 connection is torn down. Gatekeepers must monitor status for the duration of the call.	A SIP call is independent of the SIP server once established. A simple BYE command explicitly terminates a call.
Content Description	H.323 only supports H.245 to negotiate media.	SIP can use any session description format. It is not limited to only SDP.
Instant Messaging	Not supported.	Directly supported.

Both protocols reviewed so far are quite extensively supported and documented. The standards are mature enough to be widely deployed and accepted. H.323 is older and has been deployed in more mature systems. SIP, on the other hand, is very closely tied to the Internet protocols. It is fast, efficient, and, from the developer's perspective, produces tight code, which is easily readable and can produce comprehensive error codes. In the last year, SIP has overcome many hurdles and become the protocol of choice for many developers, vendors, and service providers due to its speed and efficiency.

Megaco and H.248

During the mid-1990s, Telcordia and Cisco introduced the Simple Gateway Control Protocol (SGCP). It was a single-sided model that optimized connections from traditional circuits to IP data streams. IP Device Control (IPDC) was another protocol developed by Level3, Alcatel, 3Com, Ascend, and others in a cooperative Technical Advisory Council. In the fall of 1998, features began to cross-pollinate from IPDC, and the evolution was underway.

One of the most vital points in the evolution of *H.248* or *Megaco* standards was the recognition that two distinctly different functions are performed in telephony applications. First, the call control component provides for establishing and disconnecting calls, along with cost accounting and monitoring of the network. Second, the process of media control moves the media itself. In the PSTN, these two functions are separate, with signaling and control provided by the SS7 packet network, and call traffic handled by the trunking network between COs. With the new *Media Gateway Control Protocol*, later shortened to Megaco, the concept of multimedia calling took a different tactic, splitting functionality between a *media gateway controller* (MGC) and a *trunking media gateway* (TMG). Some engineers refer to this as a *physically decomposed multimedia gateway*. This more granular approach to functionality necessitated a protocol for communicating between these two different types of gateways.

The IETF Megaco working group worked very closely with the ITU-T SG 16 group to develop Megaco/H.248, based heavily on the roots of MGCP. The design is such that the distributed system appears as a single IP telephony gateway to the outside. Megaco consists of a *call agent* and a set of gateways, including at least one *media gateway* that performs packet-to-circuit conversion, and at least one *signaling gateway*, if connected to the SS7 network and the PSTN. Megaco can also interface with SIP and H.323 compliant gateways, as shown in Table 4.2.

TABLE 4.2 Megaco Functionality			
Functional Plane	Phone switch.	Terminating entry.	H.323-conformant systems.
Signaling Plane	Signaling exchanges through SS7/ISUP.	Call agent. Internal synchronizations through Megaco and H.225/Q.931.	Signaling exchanges with the call agent through H.255/RAS
			Possible negotiation of logical channels and transmission parameters through H.245 with the call agent.
Bearer Data Transport Plane	Connection through high-speed trunk groups.	Telephony gateways.	Transmission of VoIP data using RTP directly between the H.323 station and the gateway.

Megaco describes a distributed system consisting of one or more gateways under the control of one or more call agents (CAs). One media gateway is always necessary for transporting the information. A signaling gateway is required if the IP-based Megaco network is interconnected with the PSTN. In many cases, the call agent and signaling gateway are co-located.

The call agent plays the role of providing intelligence for call control; this is distinctly separate from the packet-to-circuit conversion provided at the gateways. The CA is responsible for signaling the gateways to either set up or disconnect a call. These signals might be sent to the IP-based packet network, the circuit-based PSTN, or both networks.

The media gateway provides the mechanisms for converting audio formats from the packet media types used in IP to those used in the circuit switched network. These conversions might include functions like converting a 64-kbps PCM-encoded conversation to a more efficient encoding scheme, such as CS-ACELP, to preserve bandwidth. This gateway might be functionally spread among several devices, or have everything contained in a single hardware device. Even if distributed among multiple devices, a single CA can control the media gateway.

As shown in Figure 4.7, the call agent always communicates with the Megaco gateways. In this communication, the SA always transmits commands, and the gateway always replies with responses. The VA can communicate with the multimedia workstation or IP telephony device using a number of protocols, including H.323 and SIP.

Figure 4.7
Megaco
communications
flow.

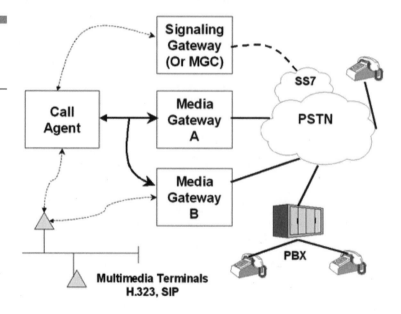

This arrangement eliminates any necessity for H.323 gateways. End users are free to use H.323 software applications, and the Megaco gateway is fully compatible for communication. The CA passes signaling requirements to the signaling gateway (or MGC) and any media-specific requirements are passed to the media gateway. The MGC signaling gateway initiates any signaling necessary to interoperate with the PSTN, while the media gateway handles message coding and transmission formatting.

Megaco Terms and Definitions

Several terms and definitions must be kept in mind when working with Megaco.

- *End points (EP)* are either physical or virtual devices that function as the source of data. Physical end points require physical termination. This might be an RJ-11 jack for a telephone, an RJ-45 jack for a computer LAN adapter, an RJ-48 jack for a DS-1 interface into a gateway, or some other physical connection. Virtual end points are defined in software. In SIP, end points are identified with an email-like address.
- *Calls* are each assigned a unique ID by the call agent. Each connection associated with a call has the same ID.
- *Call agents* provide signaling intelligence.
- *Connections* are associations set up in memory between two end points. Connections might be point to point or multipoint. Connections are maintained in memory for the duration of the call. Each connection also has a unique ID.
- *Digit maps* provide a mechanism for using regular expressions to communicate so that gateways can recognize a dialing string. These prevent the timing problems associated with sending dialed information one digit at a time.
- The *Session Description Protocol* (SDP) is used in Megaco to describe sessions. It operates in the same manner as previously described.
- The *Session Initiated Protocol* (SIP) has already been discussed. The Session Announcement Protocol (SAP) is a multicast variation that we will not explore here.

The Gateway Commands

Megaco works because call agents issue commands and gateways respond to commands. These responses may be simple result codes, and this might be combined with IP address and port number information for use by SDP.

The command structure used by Megaco is primarily simple and straightforward:

- **End point configuration**—Configures the end points or multiple end points within the gateway. This command is used to configure the end point parameters for requirements such as mu-law or A-law.
- **Notification request**—Sent to the gateway when the CA needs to be notified of some specific event. Examples of events might be the user going off hook at the telephone set, incoming fax or modem tones, or touch tone digit input.
- **Create connection**—Sent to the gateway to establish connections between endpoints.
- **Modify connection**—Modifies how the gateway handles an existing connection. This command can be used to alter the parameters of an existing call, to change the coding scheme for example.
- **Delete connection**—Signals the gateway to disconnect a call. If a call fails during the session, the gateway issues a response to the CA advising that the connections were deleted because of failures. The response to this command can also provide analytical information useful in monitoring network traffic load and performance, such as packets sent/received, packets lost, and average delay.

Megaco supports transport of these messages using UDP as the transport protocol. Megaco uses a combination of timers and counters to ensure information is delivered, because UDP is unreliable and does not assure delivery. By default, Megaco uses ports 2722 and 2427 to send messages. Multiple Megaco messages can be aggregated within a single UDP segment.

Because of the potentially tight integration between the IP network and the PSTN using Megaco, many common events in the PSTN may have to be monitored by the Megaco network devices. This is accommodated through the implementation of *event packages*. The event package describes what notifications the CA can request and what actions the gateway can generate when the CA issues a command. These event packages are described in Table 4.3. Detailed information is available in listed RFCs.

TABLE 4.3

Megaco Event
Packages

Event Package	Name	Where Defined	Description
Generic Media Package	G	RFC 2705	The generic media package events and signals can be observed on several types of endpoints, such as trunking gateways, access gateways, or residential gateways.
DTMF Package	D	RFC 2705	Defines DTMF tones.
Line Package	L	RFC 2705	The definition of the tones is as follows: dial tone, visual message waiting indicator, alerting tone, ring splash, call waiting tone, caller ID, recorder warning tone, calling card service tone, distinctive tone pattern, report on completion, and ring back on connection.
MF Package	M	RFC 2705	The definition of the MF package events includes wink, incoming seizure, return seizure, unseize circuit, and wink off.
Trunk Package	T	RFC 2705	The trunk package signal events are continuity tone, continuity test, milliwatt tones, line test, no circuit, answer supervision, and blocking.
Handset Package	H	RFC 2705	The handset emulation package is an extension of the line package, to be used when the gateway is capable of emulating a handset.
RTP Package	R	RFC 2705	Used with RTP media streams.
Network Access Server Package	N	RFC 2705	The packet arrival event is used to notify that at least one packet has been recently sent to an Internet address that is observed by an end point.
Announcement Server Package	A	RFC 2705	The announcement action is qualified by an URL name and by a set of initial parameters.
Script Package	Script	RFC 2705	Supports scripting in Java, Perl, TCL, XML, and others.

continued on next page

TABLE 4.3

Megaco Event
Packages
(continued)

Event Package	Name	Where Defined	Description
Feature Key Package	KY	RFC 3149	The feature key package groups events and signals that are associated with the additional keys that are available on business phones.
Business Phone Package	BP	RFC 3149	The business phone package groups signals other than those related to feature keys and displays.
Display XML Package	XML	RFC 3149	The XML package contains one event/signal that is used to convey XML data to and from the phone.

Call Setup Using Megaco/H.248

To step through a telephone call using Megaco, we'll have Alice call her sister Mary at home. In this case, we'll make the call a bit more complex, placing the users in different locations on the PSTN rather than directly on the IP network. IP will be used as a transport network between the telco COs. Figure 4.8 takes us through each steo of the process.

1. Betty starts the process by going off-hook and dialing Mary's home number. PSTN CO Switch #1 makes a call routing decision and seizes a trunk leading to TG1.
2. The trunk seizure and dialed digits are signaled via a Signaling System 7 message to the SS7 signaling transfer point (STP), which passes the request on to the signaling gateway.
3. The signaling gateway passes the signaling information to the call agent (CA) using an internal protocol. In many cases, these two are co-located and this might be an internal exchange of information.
4. The CA creates the trunk connection on TG 1 by sending the Create Connection command. This results in the trunk connection between PSTN CO Switch #1 and TG 1.
5. The CA performs a call routing lookup, based on the signaling information received from the signaling gateway and sends a Create Connection command to TG 2.
   a. Session properties of the two end points are exchanged at 5A in Figure 4.8.

6. This sets up the TG 1 to TG 2 connection and the TG 2 to PSTN CO Switch #2 trunk connection (shown at 8).

7. The CA informs the signaling gateway to SS7 STP in the PSTN about why the trunk was seized, and to send the called party's telephone number to PSTN CO Switch #2.

8. The call agent then issues a Modify Connection command to send the updated session parameters.

9. PTSN CO Switch #2 can now ring Mary's telephone, establishing end-to-end communications.

Figure 4.8
A Megaco telephone call, step by step.

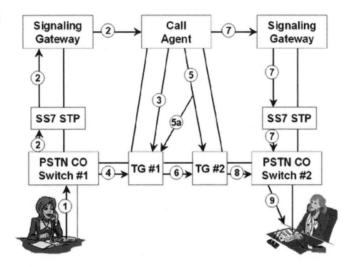

This telephone call begins with Betty at a circuit-switched connection, then traverses an IP network using Megaco, and ends at Mary's telephone, which is also a circuit-switched connection on the PSTN. Their voices are encoded using standard pulse code modulation (PCM) in the PSTN. The media gateways or trunking gateways (TG 1 and 2 in Figure 4.8 convert PCM voice into IP packetized data and back.

When implementing Megaco, different media streams and types can all be handled in the same manner. The media gateway can act not only as a trunking gateway, but it might also include multimedia conferencing equipment, a voicemail system, and an interactive voice response unit (IVR). Megaco is not restricted to IP telephony either. This approach could easily incorporate a speech-to-text conversion engine for hearing-impaired subscribers.

Real-time versus Nonreal-time Traffic

We have focused primarily on voice calls and how they get made, but it is perhaps worthwhile to take a moment to consider multimedia in general. Usually, multimedia refers to a combination of audio and video, but audio and video transmissions come in different varieties.

The voice traffic we use in the telephone network is real-time interactive voice. Two people are engaged in a conversation. Issues discussed earlier, such as delay and jitter, are quality factors that must be taken into consideration when designing the network. Videoconferencing is another real-time, two-way, interactive transmission requiring special handling in delivery.

Audio and video don't have to be real time. Broadcast television, corporate announcements, and Internet radio are good examples of multimedia traffic that is nonreal-time traffic. Although the bandwidth requirements are very similar, delivery requirements are different. If a server is sending a radio transmission over the Internet, a two- or three-second delay has little or no impact on the receiver because there is no direct interaction. Delay, while present, is not a quality factor. Several other protocols are available that incorporate buffering techniques not used in IP telephony for these services. These technologies tend to deliver information from a server, or machine, to people.

Some designers tend to think of the difference in terms of one-to-one traffic versus one-to-many traffic. This is a dangerous trap to fall into because multipoint conference calling presents a variation of one-to-many, in that the "one" might change from one person or location to another during the course of the call. It is far safer to keep in mind the concept of real-time interactive communication between people when evaluating IP telephony solutions.

Which Protocol is Needed? Which is Best?

Equipment vendors are implementing H.323, SIP, and Mecago/H.248 because the standards are mature and accepted for all three solutions. In reality, all three are necessary and play a role in one place or another. Network service providers also deploy all three.

H.323 wasn't necessarily designed for the wide area network or the Internet. As a result, it can be slow, or perceived as slow. Performance problems and scalability issues follow. It performs well in the LAN environment, which means it could provide a suitable PBX replacement or campus telephone system, but a potential bottleneck exists at the connection point to the PSTN, and great care needs to be taken in implementing interoperability support.

SIP was designed specifically by Internet developers for the wide area network. It makes the assumption that there is no intelligence in the network, whereas in the PSTN, all the intelligence resides within the network. SIP is fast and efficient, but SIP alone doesn't interoperate with the PSTN. It might be the best choice for a global corporate IP network, for internal calling from site to site.

Megaco/H.248 has a limited scope, but is defined to bring the PSTN and Internet together. Because it was designed to support both H.323 and SIP terminals, it could provide the gateway and connectivity requirements for a company to integrate the two networks. The question is whether that integration is best performed at the customer premise or at the edge of the network by the provider. Both solutions have been implemented, and both work.

The choice of protocol need not be a technical decision, but rather, a business choice. Companies that are heavily dependent on the PSTN for services may be best served by a methodical approach toward integration, using H.323 where it fits. Companies that are very Internet driven, with large IP networks, may find that SIP offers a superb solution for their internal telephony requirements. Large enterprises may well negotiate with their service providers to undertake some cooperative integration efforts that provide the benefits of a single network infrastructure for the customer, but small and mid-sized companies will likely find this approach cost prohibitive.

The key to the right choice is flexibility today and extensibility for tomorrow.

CHAPTER **5**

The Lure of IP Telephony: Can It Work for Me?

We've reviewed telephony and how voice is packetized and transmitted. We've explored issues related to quality and delivery. In short, we've spent several chapters discussing technology and learning how it works, but we haven't addressed perhaps the single most important question a business person should ask regarding any technology: "Why should I care?" In this chapter, we look at some of the offerings service providers are delivering today and explore the issues related to implementing IP telephony. We explore some of the business issues related to telephone service and the potential conversion to IP telephony.

IP telephony is one of the most talked-about technologies in the Internet sphere today. With improvements in performance and cost issues being such a large concern for business, several drivers make IP telephony an attractive alternative for business telecommunications. Because so many different business needs range from small companies with home offices to very large enterprises spread around the globe, there is no single "one size fits all" solution for telephony requirements. This should not come as a surprise, since there isn't a single solution for using existing telephony services and equipment either.

Voice versus Data

Voice, or telephony applications, requires a long holding time to support call durations, which average about 4 minutes per call. The signal is sensitive to delay and jitter because it is a real-time interactive communication. Traditionally, we've delivered this service using circuit-switched technology over the PSTN. In some cases, businesses have implemented private telephone links, such a *tie lines*, between offices to manage internal telecommunications from site to site.

Data communications are often bursty and unpredictable in nature. The durations of data transfers may be short. Traffic is often not delay sensitive because email and Web browsing are not two-way interactive services directly between people. Data services typically use all the available bandwidth for a very short duration. Data uses store-and-forward techniques to transfer across the network. Data file transfers can be very different, having very long holding times. Backup transfers of data files, large CAD/CAM drawings, and graphic files can be very large, requiring different characteristics in the network.

It isn't uncommon for businesses to have two or three completely separate networks for conducting business. Voice traffic is often handled

end of the circuit, voice traffic is digitized using traditional pulse code modulation techniques. In the packet network, the gateways might convert the data (digitized voice) into some other encoding scheme that might provide smaller sample sizes for improved efficiency. The IEC must pay the local exchange company for calls terminated in the LEC serving area. These are called *access charges* and are regulated by tariffs.

This approach has both obvious and subtle benefits, and it has been employed by a number of IP Telephony Service Providers (ITSPs):

- The use of a packet network for transport allows the sharing of resources rather than the dedicated resources required by the PSTN. This doesn't eliminate the need for traffic engineering, but changes how designers must plan for capacity.
- If the packet network used is the Internet, the service provider can now provide two necessary services from one consolidated backbone infrastructure.
- From a regulatory standpoint, the local exchange companies are heavily regulated businesses. The Internet, as an information service, is essentially unregulated, or regulated quite differently from the PSTN, which can provide a competitive advantage in pricing structure. In the PSTN, minutes of use is the standard billing mechanism, but in the Internet, flat-rate monthly billing prevails. Thus, Internet telephone calls are part of the flat rate for service, since any IP packet in the Internet is just another packet.
- Because the ISP or ITSP is not a regulated telephone company, the requirement for payment of access charges doesn't exist. The ITSP can deliver calls more cheaply, but it also means that in many cases, the local exchange company doesn't receive remuneration while its circuits are being used.
- In this particular implementation of IP telephony, the end user doesn't have to convert to IP. In fact, they needn't do anything at all and might not even know the telephone call is using IP to carry traffic from one end to the other. While the Internet is a large and rapidly growing network, there are many instances where a simple telephone will remain the device of choice for calling. Not everyone will use computers, and situations like pay phones and hotel and public phones will not need to undergo any conversion whatsoever.
- The startup cost, or barrier to entry, for an IP telephone provider is very low in comparison to the startup cost for a new telephone company using traditional PSTN technologies. IP packet switching technologies provide for greater efficiencies at a lower cost.

The real loser in this scenario may well be the local exchange company, because the major IECs are also all Internet service providers. In most cases, their ISP business is already being run as a fully separate unregulated subsidiary. This may allow them to enter the business of local service as an ISP, circumventing the requirement to pay access charges to the local exchange company. LECs are constrained from entering the long distance business without providing a substantive case to the Federal Communications Commission (FCC) regarding compliance with sharing of the local loop and other requirements of the Telecommunications Act of 1996 (TA-96). Although there have been some successes at this, notably by Verizon, it is a difficult and expensive proposition for the local telephone companies.

Certainly, equipment manufacturers have benefited from this new market. The Internet continues to grow at a rapid pace, but the addition of another IP-based technology increases the demand for products, and vendors have jumped to respond. Globally, the telecommunications services market is estimated to be over $1 trillion annually. Everyone wants a piece of the pie.

When Does It Become IP Telephony?

In later chapters, we review some specific deployments and new technologies to support them, but for the moment, let's step back and evaluate just how many different ways IP telephony is used and provided. There are a lot of options to choose from, and when some industry analyst says IP telephony will create some unfathomable billion-dollar market with some absurdly optimistic growth rate, they rarely ever delve into specifics.

We looked at one model of IP telephony in the previous section. Let's try to broaden the choices that encompass what might be called IP telephony from a business perspective.

- The gateway model described earlier certainly qualifies as IP telephony, and constitutes a large portion of the revenue being claimed today. Beyond that, the use of IP for transporting telephone traffic might be implemented by a local exchange company (LEC) within its serving area, or by an interexchange carrier (IEC), without customers ever knowing it exists.

- Pure IP telephony between computer users over the Internet is a popular communication technique that might be accomplished by users meeting in a chat room or via some instant messaging program (ICQ, AOL Instant Messenger, MSN Messenger, or the like). The users can easily choose to speak via voice, which might activate NetMeeting or some other telephony program that allows two-way voice communication over the Internet.
- Internet users might utilize some form of gateway. Two of the most notable among these are available by just clicking on www.diaplad.com and www.net2phone.com. These Web sites allow users who join and have a multimedia PC (speakers, microphone, and sound card are necessary), to place a telephone call to the PSTN for rates advertised as low as 4¢ a minute for international calls and under 3¢ a minute within the US. PC-to-PC phone calls over the Internet may be free.
- A company may elect to use PC-based software to provide internal telephone calling between employees at a single site. This LAN-based IP telephony is efficient within the network, and essentially free of charges. Calls to the PSTN might require employees to have a traditional telephone using the traditional network.
- This same company might extend that IP telephony implementation to include a gateway to the PSTN on site, processing internal calls over the LAN and converting PSTN calls through the gateway to the PSTN. This would constitute complete PBX replacement.
- A large multilocation network that implements this sort of solution might need to implement multiple gateways and sites around the country so that calls could be directed to the gateway nearest to the called party, thus minimizing toll and long distance charges. In traditional telephone networks, this was provided via PBX tie lines and *least-cost routing* or *automatic route selection* software in the PBX.

Other configurations exist. The issue may be one of semantics, but it represents a problem for business people trying to make a decision. Articles proclaim IP telephony either a whopping success or a dismal failure, but these same articles very carefully avoid specifying what succeeded and what failed in most cases. They promote generalities, and the generalities can easily balance out the equation on both sides, leaving the reader more confused than before.

All the scenarios described above are IP telephony in some form or fashion, and they all play a role in the successes and failures of the technology in the market.

It's clear that if the end user connects to the PSTN, they'll require a telephone. If they connect to the IP network, they'll need an IP device of some kind. We'll look more at IP telephones in another chapter, but IP telephony need not necessarily require subscribers to make any changes in the way they make telephone calls.

Considerations for Business Evaluations

When evaluating the business practicality of implementing any new technology, one of the greatest pitfalls is becoming enamored of the "wow factor" of the new technology. In sales parlance, this is sometimes called "selling the sizzle" rather than the "steak," It's a trap that is easy to fall into, and many businesses adopt new technologies not based on business benefits, but based on some intangible belief that the new technology will somehow improve the business. Just as dot-com companies paid a price for failing to follow standard, accepted business practices, companies that implement new technologies must accept the responsibility for performing business evaluations with due diligence to the bottom line, and not get caught up in the features and functions the new solution might provide. If a feature doesn't help the bottom line in some measurable way, it isn't a feature—it's a distractor. If it does help, it can be measured and quantified. That said, we'll now identify some ways to look at IP telephony and evaluate whether a feature adds value and favorably influences the business decision.

Cost Reduction

Will IP telephony reduce operating costs? Cost reduction can add revenue directly to the bottom line operating expense. Cost reduction can easily be weighed against the cost of implementation, and the return on investment (ROI) is easily calculated.

Cost reductions related to implementing IP telephony might bring about sweeping change, or they might be very subtle. Reduction in the cost of making phone calls is a wonderful ideal, but all factors associated with that cost have to be accounted for. Recurring and nonrecurring costs should be identified from telephone billing, service contracts, and historic data. Contractual agreements may have early withdrawal penalties that skew the time for payback on investment.

For some companies, the ability to consolidate staff is a potentially huge savings. A company that has redundant staff to support the telephone network and the voice network may be able to eliminate some staff redundancy by consolidating to a single staff. A two-fold trap that companies can fall into when making this assumption is first, the assumption that the existing IT staff can support the existing telephony requirements and one entire staff can be eliminated. This is a dangerous assumption. The IT staff is already loaded to capacity and just because IP telephony uses existing IT technology doesn't mean that no actual work is associated with support. The second danger is the assumption that because IP telephony uses IP, which the IT staff knows and understands, the IT staff won't need to learn telephony principles. Some engineering and design issues, like knowing when the busy day of the week or busy hour of the day occur, are things that telephone network administrators often understand, but are not factors in the data network. Moving telephony onto the data network moves all the issues associated with capacity planning and user support onto the data network as well. Have conservative expectations.

Increased Revenue

Will implementing IP telephony somehow increase revenue coming in to the company? How? Like cost reduction, revenue goes to the bottom line, but not always directly. Increased costs may offset revenue gains disproportionately. There may be unexpected delays and unforeseen obstacles.

Consider a Web-based company that sells goods to the retail consumer market over the Internet: the addition of IP telephony might provide a Web-enabled call center that potentially allows expanding into catalog sales and taking telephone orders. But the increase in revenue will add the cost of live customer service representatives, training, and catalog printing. Those added costs might offset the benefit derived in the first place.

Increased Efficiency

A solution that improves efficiency and reduces time to market for either goods or services provides a tangible, measurable value. Although precise efficiencies might be hard to identify, painstaking efforts should be made to identify efficiencies gained, in as much detail as possible. It is altogether too easy to speculate "solution X will make us more efficient at Y" without performing the due diligence necessary to understand whether any real value is added once the implementation is completed.

Customer Satisfaction Improvement

Every company has customers or clients. If customer satisfaction is measured and known in the current environment, will IP telephony improve it? How do you know that? If improvements can be identified and measured, they can be translated into some form of revenue equivalent.

Access to New Market Share/Market Segment

Will IP telephony provide access to a market segment or a new market share that isn't accessible using current technologies? Opening a new market can lead to revenue and a better bottom line, but it may be a circuitous and speculative path.

Provide New Service

Will IP telephony allow the company to provide a new service that cannot be provided using the current technologies? If so, what's the business case for that new service? Will it provide a competitive edge against the competition? On the other hand, if the new service means entering a new business, or changing a core competency set, make sure everyone involved has bought into the changes. Pay attention to your roots, particularly core competencies. The best staff with the purest intent to succeed cannot do a good job if the necessary core competencies aren't either present or developed.

These seem like a pretty stringent set of circumstances, but they really aren't. The target is to understand your core business. Don't invest emotionally or financially in IP telephony or any other new technology until you identify what you expect to gain. Implementing new technology just for the novelty of it is a painfully expensive proposition and counterproductive. It's an unhealthy trap that is sometimes far too easy to fall into.

The Silver Bullet

Many readers are looking for one thing, that mythical "silver bullet" that will fix their problem and make things better. Unfortunately, the old adage (credited to Robert Heinlein in *The Moon is a Harsh Mistress*) "there ain't no such thing as a free lunch" is painfully accurate.

That doesn't mean there aren't segments of the market where IP telephony makes rock-solid business sense. Several areas exist, that, upon inspection seem to provide a tight fit and a sound business case. But all business cases aren't created equal, and every situation has some unique aspects that might weight decisions in one direction or another. The bottom line is that nobody can make a good business decision without understanding all the circumstances involved. Generalities make for bad business decisions...generally. Yes, that was intentional, to see if readers are paying attention. Generalities lead to *bad* decisions. Don't rely on general recommendations on where any technology fits. Perform due diligence for your specific circumstance.

That said, based on work in the field and the telecommunications industry over many years, there are some areas where the trends indicate IP telephony is and will be a successful change. These areas have proven to fit in many, but not all, cases and seem to be the best fit today.

- **Telecom service providers**—Although this book isn't looking to telcos as a target audience, clearly the idea of using a single unified infrastructure for service delivery has merit and fits well with this group as a business approach. It can reduce cost significantly, arguably lower prices, and increase a competitive edge. So, telcos looking to improve economies may explore this area—in particular, small, independent rural telcos, which don't have behemoth organizations to maintain, may find a good fit with growth plans.
- **Internet service providers**—This is another group that really isn't the target, but a large number of small ISPs are still looking to find a competitive edge or expand their business. IP telephony may provide entry into a market that is outside typical services.
- **Large companies with multilocation networks**—This market segment often has a large private voice network deployed. Integrating voice and data into a single infrastructure can have the same impact on a large organization that it has on a telco.
- **Replace the PBX**—In a later chapter, we'll evaluate premise solutions in more depth. The PBX provides a business platform for telephony, but there is no need to maintain separate systems. PBXs are

often devices on the local area network, and they use IP as a tool for management. Why not eliminate the PBX entirely and replace it with an IP-based server that provides telephony services, essentially an IP telephony version of the PBX? This proves to be a reasonable fit for many companies, both large and small. We explore this option in more detail in another chapter.

- **International long distance**—Decreasing usage charges and improvements in cellular technology have given the United States nationwide long distance at declining rates for some time. The cost savings in cutting toll charges may not be enough for a business case. Any company that incurs high expense for high volumes of international telephone calling, however, may realize substantial savings by implementing IP telephony. Consider a small company with one office in the United States and one in the United Kingdom, but in a business that involves constant phone calls. Just moving that international telephone traffic onto the data network can pay for itself and begin saving money very quickly. (It is key to point out that not every country encourages or even allows IP telephony. If IP telephony takes revenue away from the telephone company, and the government is the telephone company, there are laws governing such use. Know your environment.)

- **Campus calling**—A company that has a high volume of internal telecom in a campus or high-rise setting can integrate voice and data and take advantage of the cost savings associated with what we would call *service convergence*.

- **Network companies**—Web-centric companies, e-commerce companies, and heavily Internet-integrated companies can create new services and offerings while taking advantage of the integration capabilities that using a single technology provides. A Web-enabled call center is far easier to implement for a Web-centric company than for one that uses minimal technology. Rather than being introduced as an intrusive technology, it provides an enabling technology.

- **Centrex users**—After nearly 30 years in the telecommunications industry, I've never been convinced that Centrex services (by whatever name they are called) are the secret "cash cow" of the industry. These expensive services are provided to help the service provider reuse what is often legacy equipment and extend its usable lifespan, thus making Centrex possibly the single most profitable sector of the entire market for telcos. There are, however, certain scenarios where these services are worth paying for and make sense. In general, the ratio of cost to value doesn't warrant using them, but everyone is reluctant to change. Many companies using these services would find a more cost effective model by implementing an IP telephony solution.

The list isn't exhaustive, nor is it intended to be. The point is to explore the options, especially if there appears to be a good fit. Review the economics on a case-by-case basis and make a sound business decision. In the chapters ahead, we review premise solutions and issues that make them untenable. We also look at network service offerings and explore how they might provide a good fit, or an interim step. For the moment, let's stick with the business issues and considerations that influence decisions.

IP Telephony in the Enterprise

A large business has a unique set of challenges to deal with. These issues may not be relevant to smaller companies, but they are certainly factors that any large enterprise must consider. Staffing and available resources are one critical difference, but another is, for lack of a better term, *agility*.

Any company develops a corporate culture that evolves over time. The more mature the company, the more entrenched the culture. The more successful a company, the greater the resistance to change may be. Changing corporate culture, processes, and practices can take time. "We've always done it this way" has become a corporate mantra that's led many businesses into a sense of false security. New approaches have to be learned through socialization over time and demonstrated by leadership. Gaining the support of the executive team (CEO, CIO, CFO, etc.) is crucial to the success of new technology deployment in any company, but a larger challenge in a larger company. A major technology shift, such as implementing IP telephony, requires a *champion*, preferably among the executive leadership.

Cross-functional teams play a necessary role in evaluating new ideas, and IP telephony is something that needs to be evaluated by every organization within the company. The IT staff can't make a good decision on solutions for operations teams unless they have a complete understanding of business needs. People who perform the jobs in the core business make up the backbone of any company. Take advantage of their knowledge of how the work gets done, and use cross-functional teams to review their requirements and perceived advantages.

Network design requirements must not be left solely to a staff of operations people. It's tempting to keep interested participants involved throughout the entire process, but do this by posting updates on the corporate intranet for everyone to view or via an email newsletter. The danger in overinvolving people in the design process is really entrusting the right motivations, but wrong skill set, with mission critical applications. If this weren't a real-world problem, it wouldn't bear mention. It's easy for people to get involved in a special project and feel a sense of proprietorship for the overall outcome. It's also easy to design problems into the solution. Make sure your design team has the right skill set and adequate resources to accomplish the task at hand. If you need to outsource to bring in the right talent, do it early in the process, rather than later. Finding design flaws later in the process is far more costly than getting an independent outside review of the plan at each stage of project development.

Change is the one constant in business. One analogy is to simplify it by using the "bicycle principle." Because almost everyone knows how to ride a bike, almost everyone understands the simplicity. Move forward or fall down. It's a simple rule, one that works. To minimize resistance, explain why changes are being made. Articulate to everyone what benefits are being gained and how the bottom line will be affected. Be open and forthcoming about business goals and expectations. If some work and staff are being reassigned, make it a good thing for everyone involved where possible. A telephone system administrator might be reassigned from an admin group to IT, but the change might involve new training, increased job skills, and enhanced opportunities for the people involved.

When planning the cutover to the new IP telephony system, do not court disaster. Do not plan for what is often referred to as a "slash cut." *Slash* may be the right descriptor, but the bleeding that often follows from this approach can be fatal. It can be tempting to anticipate a smooth cutover, with multiple sites all coming up one day and being fully functional on the new IP telephone system. It's tempting, but a cutover rarely occurs without glitches. The telephone system is the heart of mission-critical applications for most companies: imagine being without telephones for a day, two days, or a week. Plan a reasonable, phased cutover schedule that makes sense for the business, and be prepared to adapt along the way. The surest way to prevent disaster is to plan for it.

Embrace the new system. When it's up and the cutover is completed, celebrate. Let everyone know. Even if only to point out the hard work of

specialists in their specific products. Rarely have they had comprehensive training on the workings of competitive solutions, and they often don't have the proper skills to fairly evaluate your network or your needs. They may come free, and their service will be worth every penny—in other words, you generally do get what you pay for. Some companies have superb technical people, but don't place your company at the mercy of a vendor.

The second myth is that "all you have to do is add IP telephony to your network and it will work just fine." I've heard this argued by at least a dozen solution sales people in the last few months. Don't believe it. Any application, software, or device you add to your network has an impact. If a salesperson tells you to add their product so that your "voice will ride free," look to another solution provider. Once again we'll say: "there ain't no such thing as a free lunch." That will surely draw the ire of some vendors, but I've seen this done to too many customers and helped clean up the results far too frequently. To their credit, this is rarely done with purposeful intent to mislead. It's a simple case of naiveté and failure to understanding the real workings of networks.

Preparing to Implement an IP Telephony Solution—Planning for Success

A company must take several basic steps to achieve a successful cutover and implementation of a new technology. IP telephony is one of many new technologies that businesses are considering today, but the basic steps remain the same for any new technology.

- **Analysis**—Don't make the decision to move to IP telephony before identifying why you're making the shift and what the gains will be. If you decide to move ahead anyway, and some will, this analysis will help you gather risk assessment data.
- **Get the right resources**—For a larger company with a technology budget, this might mean sending staff to training beforehand or bringing in some outside help. There is no shame in admitting you need expertise from outside and working with a consulting partner.
- **Planning**—As the old saying goes, "people don't plan to fail, they fail to plan." Don't jump in blindly. Do your homework and prepare contingency plans in case things go awry.

- **Coordination**—Support your project manager. Provide the resources, get someone in a position to coordinate the overall efforts, and don't micromanage. Clearly define the expectations and requirements for the project, and then let people do their jobs. And if your project manager has a full-time responsibility for another business need, don't expect them to provide the same results as someone who's dedicated to the task at hand. That's part of the resource that you have to provide.

- **Cutover**—Be realistic, even conservative. Don't expect miracles; expect reasonable change. Don't plan for utopia, plan for glitches along the way. Have a "Plan B," in the event something misfires during the cutover. Do not anticipate a cutover that's so perfect you turn off the old phone lines on Friday evening, expecting to be up and running companywide on Monday morning using your shiny new IP telephony solution. Don't court disaster.

- **Celebrate**—When things are up and running and the new system is working, even though it may still have a few kinks, celebrate. For some companies, that may be a big party lunch. For some it might be a bonus for the team involved. For others, it will be a simple pat on the back and a thank you. Whatever is appropriate for your situation, don't fail to recognize the hard work of the people who made it happen, even if you personally are one of those people.

- **Manage the network**—Now that you're in the new environment, manage it wisely. Although we didn't talk about it in the planning phase, it's critical to think about the management and administration of a new telephone system or a data network. IP telephony is both. Don't turn it on and walk away. Maintain it and manage it. If you take care of your network, it will take care of you. If you turn it on and use it without proactive management, someday somebody is going to have to invest inordinate amounts of labor and money into unraveling the mystery it will become.

Follow these simple steps and you can assure the success of your IP telephony implementation. These are basic, common sense approaches to technology implementation. Sound business practice leads to sound business, and the whole point of implementing an IP telephony solution is to make your business healthier than before.

Does IP telephony make sense for your business? Only you can be the judge of that. But if you take the appropriate steps in gathering and analyzing the information available, you'll know your business requirements. Create reasonable expectations based on those needs and the solutions available. Manage the project for success by providing the right resources and tools for the job. Don't expect people to perform superhuman tasks, and don't expect a technology to perform miracles for your business. In the end, you'll have a reason to celebrate because you'll have a sound, reliable new tool to support your core business.

In the next chapter, we review some approaches to network performance issues that might impact your specific choice of network infrastructure or product set.

Performance, Quality of Service, and Traffic Engineering

As we delve deeper into the concepts behind IP telephony, we begin to uncover issues that may become problems because, at the core, we're trying to use technologies that have evolved in ways that they were not originally intended to go. As we explore adding voice traffic—a multimedia service—to an IP-based network, we're hard pressed to overcome one of the basic principles of networking in general: networks have always been designed to perform specific functions. Now we're attempting to use the Internet, or the IP protocol, to carry voice traffic.

This isn't a minor issue. It actually has ramifications that ripple through every facet of network engineering, in both the Internet and the PSTN. It's a new problem that wasn't anticipated. Put simply, whenever we had a task that required a form of networking, we designed a specific network to handle the task because different applications have completely different requirements.

Remember in earlier chapters, we discussed techniques for sampling and encoding voice traffic? We reviewed theories by Shannon and Nyquist, and studied how voice traffic is digitized. But to simplify it even further, let's really overlook as much of the underlying technology as we possibly can and examine very basic facts:

- Voice traffic requires long duration for a conversation. It is connection oriented.
- Voice traffic cannot experience delay. A 50 to 100 millisecond delay is about right. Voice traffic is generally referred to as *real-time traffic*.
- Voice traffic is carried in a 4-KHz voice channel, with the actual traffic being carried between 300 and 3,300 Hz.
- To convert analog voice traffic to a digital bit stream, we sample the voice at twice the maximum frequency of the channel (8,000 times per second).
- Each sample is coded into an 8-bit word.
- The 8,000 samples multiplied by 8 bits, gives us 64 Kbps of bandwidth, which matches our telephone network architecture.

These basic observations take us to the core of many technical problems we encounter today. The telephone network has been consciously designed and optimized to support voice traffic. In the telco environment, this is called *traffic engineering*, and it has been a critical component of the management and growth of the PSTN. The industry has spent years analyzing the network, watching its growth, and designing into the network every discovery possible to carry voice traffic. It is a

complex and add a variety of multimedia and streaming media components. Business automation platforms and e-commerce solutions add greater complexity to the network and increase the load on available network resources.

All of these factors must be factored together, and the performance requirements for each facet of QoS must be weighted accordingly, as shown in Figure 6.2.

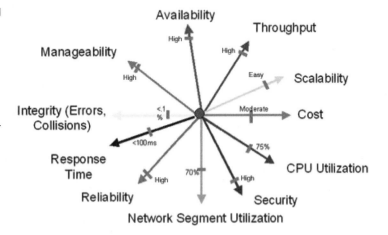

Figure 6.2
Setting the requirements Design/Operational Envelope ©2000 MultiSystems Interconnect, Inc.

Now that the network requirements have been identified, the designer can "connect the dots" and determine the *shape* of the network. As shown in Figure 6.3, this shape provides a view of network requirements that shows where heavier priorities might be required. At a conceptual level, this gives the designer a snapshot of where to expend both technical design resources and, in many cases, finances. No matter what the technical requirements of any network, cost is always a factor in business, with return on investment (ROI) being the strongest argument any network designer can make for redesign. Approaches like that demonstrated here give the network designer the information to make both quantitative and qualitative judgments about network requirements.

Documenting requirements is enough to design a new network, but there are very few businesses in operation today that don't already have a network. Readers of this book are IT and business professionals evaluating the feasibility of integrating IP telephony into an already existing environment. The very real danger, one occasionally promulgated by overzealous vendors who may use generalities, is the supposition that "the network is running fine now and will support the addition of voice with no problems." This is clearly a dangerous way of thinking, and for the network designer, it is courting disaster.

A wide variety of tools is available for comprehensive network analysis. The challenge for the designer is to measure and monitor performance of the existing production network. We cannot begin to explore the complexity of this sort of network audit within the scope of this book, and the danger is to oversimplify. Many production networks are comprised of a plethora of technologies including Ethernet, Token Ring, IP, Frame Relay, and ATM. These networks often often include multiple vendors and service providers around the globe.

Once this network audit has been completed, the network analyst can make a sound comparison between proven requirements and the documented existing network performance, as shown in Figure 6.3.

Figure 6.3
The personality of
the network
Design/Operational
Envelope ©2000
MultiSystems
Interconnect, Inc.

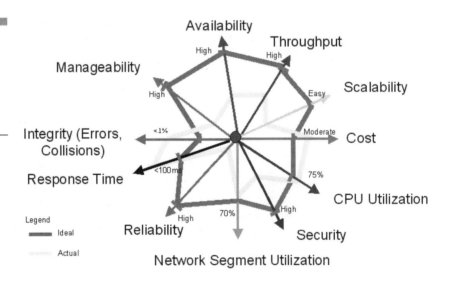

The network designer now has a visual representation of the requirements to overlay with existing conditions. This focuses on those areas of the network design that may require special attention as the new technology is implemented. This approach is useful when adding any new significant technology or application, and is often overlooked in the evolution of network technology. Too often in the business world, pressing needs, exciting opportunities, or competitive pressures dictate reactive change, leading to further and further separation between requirements and the resources available.

Acceptable IP Telephony

To get acceptable voice quality over an IP network, factors like noise, delay, echo, and jitter have to be controlled. *Delay* is assured in an IP network because of the statistical multiplexing used in routers, coupled with the bursty nature of user data. *Jitter* is the variation in delay; because different packets can easily take different paths through the network, variable delay is common. In voice traffic, jitter renders the audible signal unintelligible to the human ear, and it must be eliminated. Delay is also caused by the time required to perform the sampling we discussed earlier. Each node or router may induce *packetizing* and *routing delay* as decisions are made about how to route each packet; this is also referred to as *nodal delay*.

Certainly, a network that is not fully loaded would be expected to have lower delay. In the Internet, we have no assurances of the traffic load, and we must assume there will be congestion and delay. Thus, one way to achieve acceptable delay for voice traffic is to build a separate IP network, perhaps a private voice network. Although this solution provides a transport for IP telephony, it clearly doesn't improve progress toward complete service convergence. Although this approach may be unrealistic, it has been the approach some service providers have implemented to carry IP telephony traffic.

Quality of Service within IP

As noted earlier, IP is a best-efforts protocol that provides no guarantees. That isn't to say that the parameters we've discussed cannot be obtained using IP; however, network design becomes a crucial aspect of any implementation efforts.

In essence, all approaches to quality of service share common characteristics, although the methods of providing solutions vary widely. If a network has been properly designed, QoS is nothing more than a mechanism to implement some prioritization scheme or a routing mechanism to aggregate similar traffic types and to direct each type over a path suited to the needs of the traffic. There are many approaches with intricate differences, but every approach—save one—adds some form of overhead to the traffic to provide either a prioritization function, or a "traffic cop" function to direct traffic appropriately. The one exception, *giga-bandwidth*, is seen by many to be the ultimate solution, but this fallacy only moves the need for proper engineering out in time. We'll discuss this approach after we look at some of the options presently available.

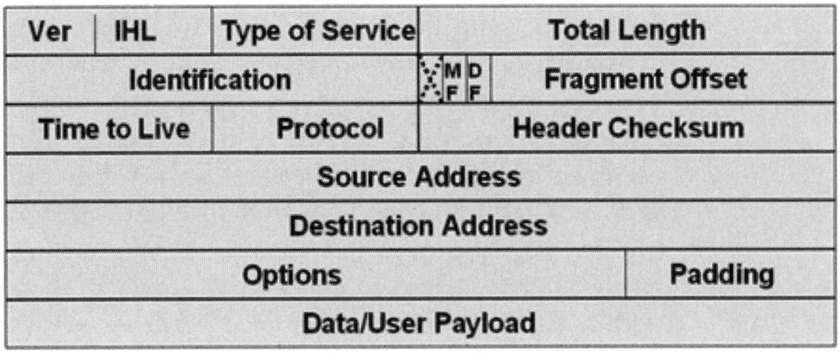

Figure 6.4
The TCP/IP packet structure.

Ver	IHL	Type of Service	Total Length	
Identification			X M D F F	Fragment Offset
Time to Live		Protocol	Header Checksum	
Source Address				
Destination Address				
Options				Padding
Data/User Payload				

We already reviewed the format of an IP packet, but it is shown again here as a reminder and also to evaluate a new Type of Service field (Figure 6.5) that can be used as a prioritization tool, and was provided in the specifications for IP.

The ToS field is one octet, or 8 bits in length. It consists of several components:

- *Precedence* is used purely as a prioritization mechanism. The first 3 bits in binary can represent a precedence value from 0 through 7. These bits provide eight possible levels of prioritization for an IP packet. The higher the precedence value, the higher priority the traffic. Note that IP still doesn't guarantee a level of acceptability, but merely a prioritization scheme.
- *Delay* indicates whether the packet requires low delay or can tolerate higher delay. A 1 indicates low delay.
- *Throughput* is a relative indicator, with a 1 indicating the need for higher throughput, or more bandwidth.
- *Reliability* is signified with a 1 to indicate that a more reliable path is needed.
- *Cost* remains generally undefined and misunderstood in use or intent.
- *Unused*. The last bit remains available for future use.

Although the designers of IP built this prioritization capability into the protocol, vendors rarely implemented or took advantage of this feature. For many years, router vendors created operating systems that didn't even read the field when processing packets, due in large part to a

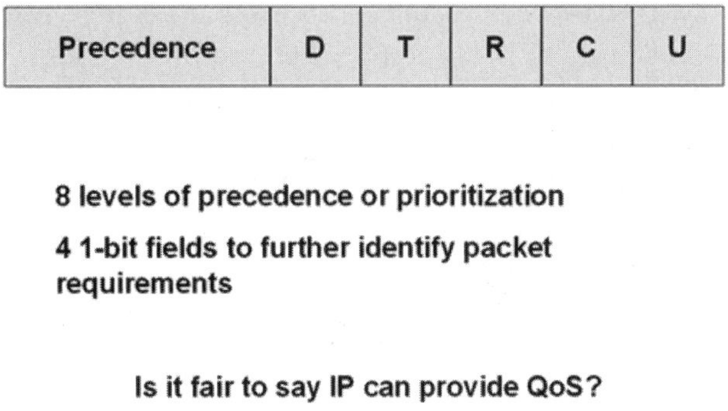

Figure 6.5
Type of Service field.

Precedence	D	T	R	C	U

8 levels of precedence or prioritization

4 1-bit fields to further identify packet requirements

Is it fair to say IP can provide QoS?

lack of standardization. The Internet Engineering Task Force (IETF), the IP standards body, has never assigned values to any of the subfields in the ToS field, thus a precedence of 0 might be the highest precedence to one vendor, and the lowest to another. Real-world implementations vary greatly in the absence of a standardized approach.

We will not explore routing protocols like *Routing Information Protocol* (RIP) or *Open Shortest Path First* (OSPF), but these routing protocols are built to operate on a single metric—the "best path" approach. Since the best path for voice might be different from the best path for Web traffic, these protocols require redesign to support multiple path selection metrics.

Given that IP provides no guarantees of any kind of delivery or prioritization, if a user needs to specify some particular network requirement, or QoS, the user must add something to IP to provide it. Given the real-world implementations of IP, confusion arises. Many companies have implemented their own private, internal IP network, and many different ISPs have not all taken the same approach to providing QoS. Given the variables, we now explore some of the different approaches that are in use, and evaluate some strengths and weaknesses.

Three QoS Philosophies for IP

Three methods are used to implement QoS in IP:

- **Signaled quality of service**—as implemented by *Integrated Services* (IntServ) requires the addition of a signaling protocol to IP. In this approach, a user application sends a call setup signal to the network requesting a particular set of service delivery parameters. Ideally, the network responds with the necessary resources to deliver traffic according to the application requirements.

- **Provisioned quality of service** as implemented by *Differentiated Services* (DiffServ) requires that specific paths through the network be available for different classes of traffic. These paths might be provisioned as part of the network design, or set up on demand. This approach generally looks to aggregate similar traffic types onto the same paths.
- **Bypass or "shim" quality of service**—as implemented by *Multiprotocol Label Switching* (MPLS) is an approach that circumvents hop-by-hop routing in IP, which can dramatically improve nodal processing time. MPLS is fully compatible with both Frame Relay and ATM, and it has become very popular for a number of uses.

Integrated Services (IntServ)

The Internet Engineering Task Force (IETF) created a suite of protocols that are used together and referred to as Integrated Services or IntServ. The IntServ working group charter and information is available at www.ietf.org/html.charters/intserv-charter.html. The objective of this working group is to standardize the requirements for a service model that can transport audio, video, real-time, and classic data traffic over single IP network architecture. This suite is defined by a series of RFCs:

- **RFC 2210**—The Use of RSVP with IETF Integrated Services
- **RFC 2211**—Specification of the Controlled-Load Network Element Service
- **RFC 2212**—Specification of Guaranteed Quality of Service
- **RFC 2213**—Integrated Services Management Information Base Using SMIv2
- **RFC 2214**—Integrated Services Management Information Base Guaranteed Service Extensions Using SMIv2
- **RFC 2215**—General Characterization Parameters for Integrated Service Network Elements
- **RFC 2216**—Network Element Service Specification Template
- **RFC 3006**—Integrated Services in the Presence of Compressible Flows

Under IntServ operations, users request a particular type of network service using *Resource Reservation Protocol* (RSVP). RSVP has two key aspects, *policy control* and *admission control*. Policy control is used to determine if the user has authorization to request the specified resources from the network. Assuming the user has the necessary permission, admission control governs the process for allocating and reserving the requested resources, or setting up the path for the connection. As we'll see, RSVP passes the user requirements from node to node through the network requesting resources. If any node along the way is unable to comply, the request is denied and no connection is established.

Real-time Transport Protocol (RTP) is used to assure the integrity of a session with time stamping and the sequencing of UDP segments. *Real Time Control Protocol* (RTCP) is used in conjunction with RTP and provides some level of monitoring capability.

Although IP theoretically has a rather comprehensive prioritization scheme available, IntServ provides three levels of prioritization. They are:

- **Best effort**—which is exactly what IP provides in existing networks.
- **Controlled load**—which is described in the standard as providing "data flow with a quality of service closely approximating the QoS that same flow would receive from an unloaded network element."
- **Guaranteed service**—which provides the higher level of assurance that might be required for real-time traffic like voice or video.

Resource Reservation Protocol

RFC 2205 defined the first version of RSVP to provide the necessary signaling capability that allows a user application to send a request to the network requesting a particular set of requirements for a given session. This set of requirements is referred to as the *template*. (We keep referring to user applications rather than users, because it's important to understand that the person using the computer will not be required to understand QoS or application requirements. The applications will have requirements coded in at acceptable levels to provide satisfactory service.)

RSVP allows user applications to identify three of the parameters that might be necessary for a given application

- Throughput or bandwidth
- Delay
- Jitter

In Figure 6.6, the user application makes a request of the network. RSVP *policy control* evaluates the request against a permissions table to confirm that the user has the necessary permission to reserve the requested resources. If that passes, RSVP then engages *admission control* to determine whether or not the network has appropriate resources available. The network passes the template along the network from node to node in a PATH message until it reaches the receiver. Each node must run an *RSVP daemon* to validate the request for QoS. In simple form, this specification is a request, passed from router to router, identifying a need. If an intermediate router can meet the need, the request is passed to the next node. If not, the request is denied. Intermediate nodes play a crucial role in RSVP, because each intermediate node must be able to provide the necessary QoS PATH requested in the template.

Figure 6.6
The user application makes a request of the network.

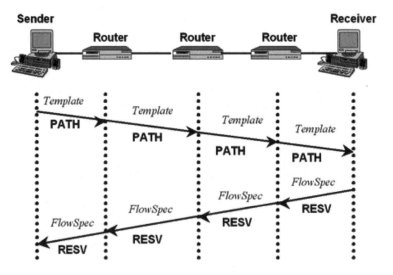

Once a path has been identified, each node must pass a *flow specification* in an RSVP message back along the network from node to node. Since we've gone to the trouble of reserving resources across the network, we now must ensure that we use the resource allocated. This *flowspec* reservation is each router's way of knowing where the next router in the reserved path is to pass packets associated with the call or session.

Clearly, a number of issues are involved in implementing RSVP for QoS in the Internet:

- **Overhead intensive**—Call setup isn't something IP was designed for, and the addition of a signaling protocol adds to the processing required at each router in the path.
- **Routing protocols like RIP and OSPF only support a single metric**—RSVP does not provide any solution for this issue, implying that, in order to be successful, the network must be overengineered beyond "normal" carrying capacity.
- **RSVP scales very well in the multicast environment (one-to-many transmissions), but does not scale well for unicast (one-to-one) traffic**—Since telephony is primarily a one-to-one connection, scalability quickly becomes a problem in a large network and is unattainable in a network like the Internet.
- **Since intermediate nodes must be able to comply with requests, every router in the Internet must be either upgraded to support RSVP or replaced.** This would require universal support and acceptance that has not developed. Routers have to maintain state tables containing information about every session. The processing load and memory requirements could drive the cost of routers up significantly.
- **RSVP doesn't actually provide any QoS, but only a technique for requesting specific requirements of the network.** It is a signaling protocol that still requires other methods for implementing QoS.

Many network engineers concur that IntServ and RSVP can provide a solution, but only in a network that is completely under the user's or provider's control. In the Internet, data traverses many networks en route to its destination. IntServ can work well in a fully controlled environment, like a customer-owned private network. It may also prove useful at the edges of the Internet, either in customer networks or the local metropolitan portion of the Internet provider's network.

Internet service providers have almost universally declined to embrace IntServ as a solution to the QoS problem. The value provided has not proved to be something that can overcome the cost of conversion and implementation. IntServ is finding some application in private networks and continues to be supported by many vendors. *Resource Reservation Protocol for Traffic Engineering* (RSVP-TE), a variation, has been deployed with success in conjunction with *Multiprotocol Label Switching* (MPLS).

Differentiated Service (DiffServ)

Differentiated Services (DiffServ) is also a development of the IETF. The DiffServ working group charter and information is available at www.ietf.org/html.charters/DiffServ-charter.html. The objective of this working group is to employ "a small, well-defined set of building blocks from which a variety of aggregate behaviors may be built." This suite is also defined by a series of RFCs:

- **RFC 2386**—Per Hop Behavior Identification Codes
- **RFC 2474**—Definition of the Differentiated Services Field (DS Field) in the IPv4 and IPv6 Headers
- **RFC 2475**—An Architecture for Differentiated Services
- **RFC 2597**—Assured Forwarding PHB Group
- **RFC 2598**—An Expedited Forwarding PHB
- **RFC 2983**—Differentiated Services and Tunnels
- **RFC 3086**—Definition of Differentiated Services per Domain Behaviors and Rules for Their Specification
- **RFC 3140**—Per Hop Behavior Identification Codes
- **RFC 3246**—An Expedited Forwarding PHB
- **RFC 3247**—Supplemental Information for the New Definition of the EF PHB
- **RFC 3248**—A Delay-Bound Alternative Revision of RFC 2598
- **RFC 3260**—New Terminology and Clarification for DiffServ

We do not evaluate IPv6 in this book, but we do feel its impact at times. In the development of the next-generation IP protocol, the ToS field was the subject of much debate. In IPv6, the protocol has been expanded to provide better support for delivering QoS. In the IPv6 packet, a field referred to as the DiffServ field replaces the current ToS field. It uses 6 bits of this field as the *Differentiated Services Codepoint* (DSCP) to identify how the nodes in the network should handle each packet. Routers handle packets based on a set of forwarding treatments or *Per-Hop Behaviors* (PHB). These rules must be predefined in each network element or node. This is the *provisioned approach* to QoS in the network.

The DiffServ approach categorizes traffic into different classes of service (CoS). Like services are then aggregated for similar handling, and routed over paths configured to support that particular CoS. Packets are classified at the edge of the network, perhaps at the ingress router to the Internet, then the designated forwarding treatment for that CoS is applied. This approach results in a much coarser granularity at each individual node, thus reducing the requirement for large state tables and increased processing power.

DiffServ has two primary components, described in RFC 2475:

- **Packet marking redefines the packet's ToS field and uses 6 bits as a coding scheme to classify packets into a CoS**—The use of 6 bits provides a prioritization scheme that can identify 64 different types of traffic or aggregates.
- **Per-hop behaviors govern how an individual class or aggregate is handled**—This is defined via *behavior aggregates*. In essence, the PHB describes the scheduling, queuing, and traffic shaping policies used at a specific node for routing the traffic.

DiffServ scales to very large network size and has been implemented by a large number of equipment vendors. It is widely supported and has been deployed by several large service providers. It is far more complex than discussed here, but information is readily available on the Web to provide readers with great specific technical detail.

Multiprotocol Label Switching (MPLS)

Multiprotocol Label Switching (MPLS) is yet another development of the IETF. The MPLS working group charter and information is available at www.ietf.org/html.charters/mpls-charter.html. The objective of this working group is "standardizing a base technology for using label switching and for the implementation of label-switched paths over various link-level technologies, such as Packet-over-SONET, Frame Relay, ATM, and LAN technologies (e.g., all forms of Ethernet, Token Ring, etc.). This includes procedures and protocols for the distribution of labels between routers, encapsulations, and multicast considerations." As with all IETF projects, this suite is also defined by a series of RFCs:

- **RFC 2702**—Requirements for Traffic Engineering Over MPLS
- **RFC 3031**—Multiprotocol Label Switching Architecture
- **RFC 3032**—MPLS Label Stack Encoding
- **RFC 3034**—Use of Label Switching on Frame Relay Networks Specification
- **RFC 3035**—MPLS Using LDP and ATM VC Switching
- **RFC 3036**—LDP Specification

- **RFC 3037**—LDP Applicability
- **RFC 3038**—VCID Notification over ATM link for LDP
- **RFC 3033**—The Assignment of the Information Field and Protocol Identifier in the Q.2941 Generic Identifier and Q.2957 User-to-user Signaling for the Internet Protocol
- **RFC 3063**—MPLS Loop Prevention Mechanism
- **RFC 3107**—Carrying Label Information in BGP-4
- **RFC 3209**—RSVP-TE: Extensions to RSVP for LSP Tunnels
- **RFC 3210**—Applicability Statement for Extensions to RSVP for LSP-Tunnels
- **RFC 3212**—Constraint-Based LSP Setup Using LDP
- **RFC 3213**—Applicability Statement for CR-LDP
- **RFC 3214**—LSP Modification Using CR-LDP
- **RFC 3215**—LDP State Machine

An excellent MPLS Resource Center is available online at www.mplsrc.com.

MPLS is sometimes referred to as a "shim" protocol because it adds a tag in front of the header of each packet entering the network. This approach was deployed by Ipsilon, now part of Nokia, under the name *IP flows*. Cisco Systems also deploys a sophisticated packet tagging approach known as *tag switching*. MPLS is an outgrowth of the success of these efforts.

To understand how MPLS operates, several terms need some explanation:

- *MPLS* is used to refer to the protocol itself, and that's how we use it in the book. In the rapidly evolving optical networking environment, it is sometimes also used to mean *multiprotocol lambda switching*.
- *LER* is a *label edge router* and is the device at the edge, or ingress and egress, of an MPLS domain.
- *LSR* is a *label switching router* (sometimes called a label swapping router). Any device capable of supporting MPLS is an LSR. This might include ATM switches, routers, Frame Relay switches, or other network devices.
- *LSP* is a *label switched path*, a stable path followed by MPLS traffic.
- *FEC* is a *forwarding equivalence class* or aggregate of packets that are all treated in the same manner.
- *LDP* is the *label distribution protocol*, or method network nodes use to exchange information about the labels used.

MPLS supports the widely accepted use of IP, and incorporates mechanisms for integrating some Layer 2 traffic management and prioritization methods from technologies like Frame Relay and ATM. Routers at the edge of the network (LER) apply a label to each packet. In the case of IP, this label is a shim header inserted into the IP packet. Rather than hop-by-hop routing decisions at each node along the path, packets are switched onto a label-switched path using Layer 2 technologies. This minimizes the processing required at the intermediate nodes, thus allowing for faster packet transfer because the IP header is no longer processed at each node. At the destination edge of the network, the label or "shim" is removed, and the packet is delivered to the recipient.

Figure 6.7
MPLS architecture overview.

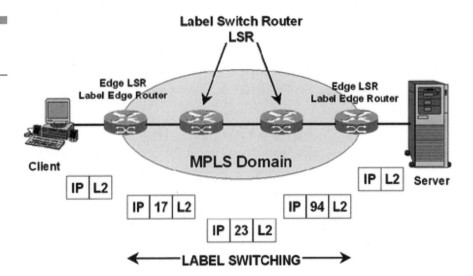

In Figure 6.7, we see that the label "17" is applied to packets entering the MPLS domain from the client workstation. The label "17" only has significance between the LER on the left and the LSR on the left. The LSR switches that label to "23," which then has significance between the two LSRs. The LSR on the right switches the label to "94" as it passes the data along to the egress router on the far right. Using this technique, the LSRs can switch traffic at Layer 2, rather than make routing decisions at Layer 3, thus improving performance and providing for optimal paths. Information about the labels and aggregate traffic requirements is shared between devices using a separate Label Distribution Protocol (LDP).

Implementing MPLS into an existing network is a very complex task, but due to its approach, it can be used not only for traffic prioritization, but also for virtual private networks (VPNs). In effect, it is a very useful traffic shaping technology that is receiving widespread support from numerous service providers and all of the major equipment vendors. (Cost and complexity are issues, and many smaller Internet providers are not able to convert their networks to newer technologies as quickly as the large nationwide providers.)

MPLS has received a great deal of recognition for its ability to integrate IP traffic with the QoS mechanisms built into ATM, which is widely deployed in the Internet backbone. Because MPLS labels can be correlated directly to the traffic classes provided in ATM, it presents a convenient blending of the two technologies, although any other Layer 2 technology can be used. MPLS is widely used in variations developed specifically for optical networking.

MPLS is not a true QoS mechanism, since some other protocol must be used to select the "best path" for any given type of traffic.

Where's the "Quality"?

A student in an IP telephony class once asked a most telling question: "Where does the actual quality get improved?" This question really bears some consideration:

- **IntServ adds the overhead of RSVP and is used to reserve a path**—IntServ itself does nothing to assure that quality is available, and if the network cannot comply, RSVP reservations are denied.
- **DiffServ uses the ToS field in an IP header to aggregate traffic into classes of service (CoS)**—But DiffServ does nothing to provide a path through the network that meets those requirements.
- **MPLS adds the benefit of bypassing the overhead of a routing decision at each intermediate node**—But MPLS does nothing to actually ensure that appropriate paths are available.

Each approach really boils down to being a prioritization and aggregation scheme that allows for either requesting resources and hoping they're available or relying on some other protocol or method for providing the paths and resources necessary. These QoS approaches provide traffic management and, in some cases, traffic shaping capability, but they don't directly improve the quality of the network. In truth, they each add overhead to the network in some form. So where is the quality?

Gigabandwidth

We've mentioned optical networking several times. One approach to providing quality to a network is to dramatically increase the carrying capacity or bandwidth of the network. Many pundits and technologists have argued that with the exponential increases in optical networking, we have achieved a glut of bandwidth. Some have even proclaimed that "bandwidth is free"—although we all seem to be paying a higher price than we'd like to pay.

Nobody disputes that bandwidth is getting cheaper. And the advances in optical technologies are astonishing: a strand of fiber can carry far more information than a 500-pair bundle of traditional twisted pair wire. And the advances continue at a rapid pace. So, if bandwidth is free, or so commoditized as to have a negligible cost, are QoS mechanisms even necessary?

One of the early drivers of IP telephony was to move voice traffic off the more expensive PSTN onto a cheaper network, like the Internet. While this presents an intriguing argument, it also presents a true "apples to oranges" comparison. The PSTN and Internet are regulated, managed, and priced under completely different methods. The PSTN is a highly regulated environment, while the Internet remains essentially an unregulated data service.

Optical networking isn't the only technological advance. DSL deployments have proven far more successful as providers gained experience. While still not deployed to a very large portion of the population, high-speed access to the Internet has become more readily available. Ethernet technologies have made Gigabit Ethernet a far more cost-effective and common technology than we would have believed a few years ago.

Will we ever see end-to-end gigabandwidth? Perhaps; but given the real technical challenges of existing local loop technologies, coupled with the challenge of delivering massive broadband connectivity to rural areas, this solution is still many years away.

Another argument is that providing a user with gigabandwidth is merely a temporary patch. The Internet has done more to encourage consumption of data that its early designers imagined. End users always find creative uses for technology that the inventors or designers never dreamed of. Today, the idea of a gigabit data stream to a residence seems to eliminate any need for quality of service.

The question becomes whether or not some traffic is more critical than other traffic. For many businesses, telephone traffic is the most mission-critical application there is. A call center that takes catalog orders or makes airline reservations cannot function without telephony. Assigning a lower priority to email and Web traffic is a sound business decision for such a company, but to do that, some prioritization scheme is necessary, whether it's one reviewed here or some entirely new approach.

The best solution to QoS issues involves the well-being of the entire network. Processing power must improve. Bandwidth must increase. Traffic engineering techniques must evolve. Prioritization for mission-critical traffic must be implemented. No single "silver bullet" solution can meet all the demands placed on IP and the Internet today, because the problem is not one single problem. Only a complex solution will effectively address the myriad performance concerns in today's networks.

CHAPTER **7**

Gateways between Networks

Earlier we discussed the use of an H.323 gateway to connect the IP network to the PSTN. Several variations in gateways exist, but they all are used any time we need to connect from one kind of a network to another kind of network. Gateways perform conversion in coding, protocols, transmission technologies, and physical connectivity.

Why Gateways Are Necessary

Gateways are perhaps the key differentiation that provides the transition between "voice over IP" and "IP telephony." Many people use those two phrases interchangeably, but the implications are quite different.

Voice over IP (VoIP) has been achievable for several years. Put simply, it's just packetizing the voice signal and using IP as a protocol to transmit the information from one place to another. IP is unaware of the payload contained in packets. Anything that can be converted into a string of bits can be carried in IP.

IP telephony implies much more. Telephony, as we've discovered, includes a rich and robust set of features and functions that have evolved over time. Telephony is more than just transmitting voice from one place to another. It involves the ability to set up and tear down connections. Accounting capability is a requirement in the telco environment, as are added features like call waiting, conference calling, or music on hold, which add to the rich set of features that comprise telephony.

At inception, the IP telephony gateway was simply a conversion tool to connect different types of networks. Early solutions provided for packetized voice using IP within the local area network environment. This solution makes perfect sense, because LANs are deployed in many businesses. The available bandwidth has grown from a shared 10 Mbps to a switch topology, then to 100 Mbps and beyond. Ample bandwidth and network capacity allows for not only voice transmission, but premise-based video in the LAN.

Telephony, however, as a voice application has characteristics that are different from video. It's ubiquitous: the PSTN provides service everywhere, and for a telephone application to be effective, users have to be able to make phone calls not just within the LAN. Internal calling in a closed environment may have some limited application, but the ability to call customers, business partners, or home quickly is a necessity for any application to be adopted. Gateways are used to provide different feature sets, but all with one common purpose: connecting multiple networks.

Earlier, we reviewed some of the protocols used in IP telephony. H.323 employed the concept of the gatekeeper and gateway as mechanisms for managing the network. Those devices have converged into a single gateway in many instances today. This newer type of multifunctional gateway is called a *softswitch* by most vendors. In the case of Megaco or H.248, we saw two gateways, a signaling gateway to connect to the SS7 network in the PSTN and a trunking gateway to connect the actual paths that carry the voice traffic. This approach of separating functions into "command/control" and "traffic-bearing" components is being widely adopted for both efficiency and economy of scale in a number of technologies today. Since the two functions are distinctly separate, they may be bundled, or they may be discrete devices, depending on a particular vendor's design approach.

Because gateways can be deployed in several different models, we'll explore some of the alternatives and look at where and how they've been used. In some of these approaches, either an ISP or telephone company deploys the gateway because technology markets are very competitive and providers are constantly searching for new services or more cost-effective delivery strategies. In other cases the customer, on site at the premise, deploys that gateway. This approach may afford features not available through a provider, but the customer must take on some responsibility for the technology. We'll explore the customer gateway in more detail in the premise solutions chapter.

The Service Provider Basic Model

We've already seen the basic model used by service providers, but this model has been adapted to fit a wide variety of different business situations. Depending who owns the gateway, this same physical model can be used by the incumbent telephone provider, the interexchange carrier, an ISP, or a new Internet telephony service provider (ITSP). Functionally, the approach is universal, and business management or service approaches provide the variations in many cases.

Perhaps the simplest approach involves the placement of gateways in the telco CO. This approach provides three distinct service provider models.

The ILEC Gateway

The telco can be the provider of the gateway by adding equipment to the existing CO infrastructure. This can be implemented in two ways: the first approach is to add the hardware gateway to the network. This approach is shown on the left side of Figure 7.1. Another option for the local telco is to upgrade the CO switch to incorporate IP telephony support into the switch itself. These upgrades are readily available for large CO switches from companies like Lucent Technologies and Nortel Networks, but also in switches used by smaller telephone companies. The advantage to this approach is tighter integration of the two technologies and a more unified management and administration because of the single vendor solution.

This method allows the addition of IP telephony functionality on a CO-by-CO basis, providing an approach the permits the telco to scale the network over time. As more COs support IP telephony, the reliance on the PSTN infrastructure within the cloud can diminish over time. With advances in IP networking, optical transport, and high-speed networks, raw capacity may be cheaper to provision within the Internet suite of technologies than in the traditional telephone network.

Figure 7.1
The ILEC VoIP
gateway.

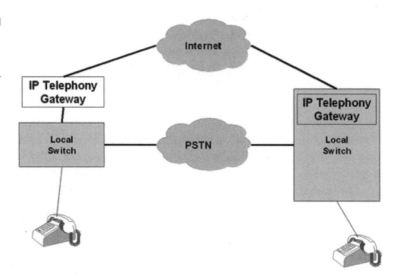

Is this IP telephony? The consumer isn't affected in any way. Nothing changes on the customer premise, and the customer may not even realize IP is being used to transport telephone calls. However, IP is carrying the telephony traffic, and the revenues associated with this model are often included in reports of IP telephony revenues. It is conceivable that the telco could implement the changes and convert the network to IP, and the only change that customers would see is reduced cost as the network becomes more efficient. Certainly, the convergence to a single infrastructure provides some cost benefit to the telco over time, but time is the problem. This is not a solution that telcos can or will deploy overnight, because there isn't a driving demand for the deployment—it offers no value to the end user. The only thing it might do is incrementally lower expense. Most telcos are exploring this approach, and many trials have been implemented, but successes have been small.

Certainly, this approach may lead to a new paradigm for the future. How do I pick up my telephone at home and dial a number that "rings" the computer of an Internet-connected user in some other location? This connectivity isn't widely available, but as the Internet becomes a more integral part of our daily lives, the need for this ability is clear.

The ITSP Gateway

Oddly enough, the physical model used by ITSP is identical. The differences are business issues rather than technical ones. Since the breakup of the Bell System in 1984, there have been numerous efforts by companies to enter the telecommunications market. Some have been very successful; some have struggled.

In the ITSP model, the service provider maintains an IP backbone and must connect to the PSTN at a telco CO. The ITSP will often negotiate a leasing arrangement for floor space within the telco CO. This colocation arrangement provides a secured cage in the CO where the ITSP can install equipment and gain easy access to the telco network and equipment. This model is a result of several requirements of the Telecommunications Act of 1996 mandating that the incumbent LEC provide freer access to its local customers. Because this requirement is an important consideration to an ILEC trying to gain approval for entering the long distance market, some ILECs have worked hard to provide this type of access.

The ITSP places the gateway in or near the telco central office. The ITSP then must purchase some trunking connectivity into the CO. To be cost effective, this means the provider must place a gateway in every local CO it wishes to serve, or at least in all the calling areas.

IP telephony has often been referred to as a disruptive technology, and this model provides a clear example of how it could disrupt the telecommunications industry. The ITSP is often an ISP that is expanding into new service offerings. Given all the advances in optical networking, providers like Qwest and Level3 certainly have the network capacity to carry additional services. Many Internet providers have added fiber backbone infrastructure and increased capacity in the past few years. In the Internet, even at peak times, the capacity is rarely used. For an Internet provider, it only makes sound business sense to find additional markets for the capacity that is so readily available. Why not enter the telephony market, but without the constraints of telecom regulation that the ILECs endure? The ITSP model is shown in Figure 7.2.

Figure 7.2
The ITSP VoIP
gateway.

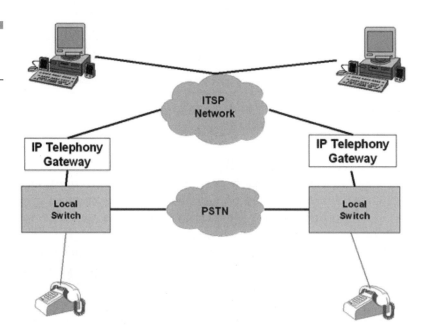

Selling surplus capacity by delivering a new service brings in new revenue from existing investments and makes for a pretty sound business model—expansion of offerings without substantial capital investment. Of course, there's investment required in gateways to provide the necessary support to integrate new solutions into the existing infrastructure, but it's exponentially less than building an entire new network. It has the advantage of being IP based, which is already the core protocol in the underlying networks.

dynamic database that allows a static telephone number to correlate to a dynamic IP address every time the user connects.

While convergence presents many wonderful opportunities to bring services together onto a single infrastructure, it also brings several added degrees of complexity to the network. The existing IT staff may not have a deep enough understanding of telephony requirements. The telecom staff may not be familiar with data networking technologies. The skill set required to design and manage this network is a blended skill set that has not been necessary in the past when the two networks remained distinctly separate. As networks and services converge, the pool of talent with the right skills to design, manage, and maintain these networks shrinks. It's important to ensure that the support staff receives the proper tools and training to keep these mission-critical services operating at acceptable performance levels.

Types of Gateways

We've identified several different types of gateways so far, and we're going to add one more variation before we move on and bring up one more standard. The Generic Requirement-303, or GR303 standard, was developed by Telcordia (formerly BellCore) and is mostly used in North America. The rest of the world uses a V5.2 standard from the European Telecommunications Standards Institute (ETSI). These specifications describe how a subscriber connects to a Class 5 telco CO switch using a digital loop carrier. Subscribers connect to Class 5 switches in the PSTN. Class 3 and 4 switches are interconnection and trunking switches within the network itself. GR303 describes differences in connectivity for things like loop start and ground start circuits, which we won't explore here.

One factor that has made GR303 such a critical standard is that with the provision of the Telecommunications Act of 1996 (TA-96), the PSTN was opened up to allow *competitive local exchange carriers* (CLECs), to access the local loop, which is owned by the incumbent carriers or ILECs. GR303 provides a standard set of operating rules that these carriers can use in connecting to CO switches in the PSTN.

Gateways that connect into a Class 5 switch are now often both GR303 and V5.2 compliant to ensure maximum compatibility. Many designers view GR303 compatibility as a crucial step in the migration from a time division multiplexing-based (TDM) network environment to the packet-switched environment of IP, Frame Relay, ATM, and DSL. GR303 allows gateways to provide functions like POTS, pay phone support, and IDSN services while leveraging telco investments in fiber-connected digital loop carriers. Although GR303 is a standard, major vendors each implement it in their own special way. As is often the case with standards, compliance and interoperability don't necessarily mean the same thing.

In the United States, as the industries continue to evolve, the lines of distinction between providers blur to the point of utter confusion. Local exchange carriers have entered the long distance market. Interexchange carriers obtain CLEC status in order to provide local services. Both have created Internet services and now operate in some combination of regulated and unregulated entities or subsidiaries of the same corporation. CLECs entered the market to provide data services, but quickly migrated to data, notably in DSL services, becoming data local exchange carriers or DLECs. Companies that were pure Internet providers expanded into telephony and became CLECs. AT&T bought heavily into the cable television industry with TCI, and moved into cable modems and yet another delivery model with a different local loop technology. To make matters more confusing, many of the players moved into the cellular or PCS market for telephony, and several moved into fixed wireless delivery of data using multichannel multipoint distribution system (MMDS) or local multipoint distribution system (LMDS), too.

One thing is clear; everybody wants a piece of the action, wherever the action might be. As a result, standards have become more important than ever in allowing for interoperability between vendor equipment deployed by the multitude of providers available.

Gateway or Softswitch?

We've used the term gateway throughout the text to this point, but gateways have also evolved as IP telephony has become a more viable technology. Gateways initially provided only the mechanism for converting information from one type of network to another.

As you may recall, in the H.323 standards, there were both a gateway and a device referred to as a gatekeeper. The gateway provided call control functions, supporting the conversions necessary to communicate between the IP network and the PSTN. The gatekeeper provided three primary functions: registration of users to participate in the network, admission of users based on the permissions assigned to their accounts, and status functions for tracking call progress through the network.

We haven't yet addressed one function upon which the PSTN relies, and that's billing. Without a billing mechanism, the PSTN would function, but nobody would ever be paid. In the IP network, billing is perhaps better thought of as an auditing and accounting function. Without the ability to maintain audit trails and account for traffic, billing becomes a problem. Traffic engineering without some audit and accounting mechanism can only track IP packets. The ability to measure the impact of voice traffic on the network requires a separate audit mechanism to measure that traffic.

These devices have all been implemented in a variety of different systems. Some systems were simply computers with customer software running on a standard hardware platform, usually a server-class machine. Others were built on proprietary hardware platforms optimized to perform special functions.

As the individual functions matured and evolved, equipment vendors began to consolidate them into a single device. To differentiate the device itself, and incorporate a variety of features from a range of other products, the term softswitch was used to describe this new hybrid device. A softswitch is sometimes defined as software using open standards that can perform distributed communications functions on an open computing platform.

We might divide IP telephony into three different features:

- **Applications**—Includes functions such as call accounting auditing, user administration, and an application programmers interface (API) to interconnect with the advanced intelligent network (AIN) of the PSTN.
- **Control**—Functions that denote features like call control and signaling.
- **Service Delivery**—Provides the actual transport and delivery of the payload information.

In this model, the softswitch encompasses the applications and control features, whereas service delivery is the responsibility of the underlying IP network. This model allows for the extensibility of adding new features as technologies continue to mature, and expands the IP telephony model to that shown in Figure 7.4. Through the standard communications of SS7, the softswitch is now able to integrate the enhanced telephony features of the advanced intelligent network (AIN).

Figure 7.4
Softswitch integration
between IP and the
PSTN.

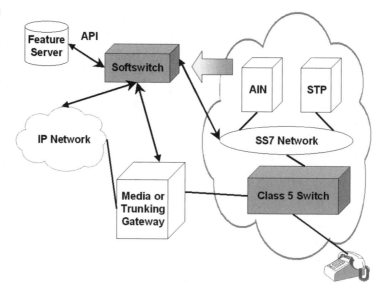

Figure 7.4
Softswitch integration between IP and the PSTN.

Many of the market drivers for this evolution of gateways and softswitches are the same drivers as for IP telephony in general:

■ Two big vendors provide circuit switching equipment in North America, Nortel and Lucent. Although other vendors exist, these two have a nearly de facto market monopoly among the incumbent ILECs. Features are often proprietary.

■ Circuit switches are vertically integrated by vendor and have become too large and expensive to maintain and upgrade. Adding features is a major undertaking.

■ Maintaining and growing two networks is expensive. Providers are looking for the economy of scale that using one infrastructure can provide.
  – Although voice traffic has been the primary source of revenue, data traffic provides the primary consumer of bandwidth.
  – The voice market is still growing.

■ Packetizing voice can leverage unused capacity in data networks.

■ IP can bring new features to bear that are not readily available in the circuit-switched environment.

The integration of networks provided by gateways and softswitches provides a boon to service providers. In some cases, it allows the same service to be offered at a reduced cost. In other cases, it stimulates revenue generation from an existing infrastructure. For equipment vendors, it provides a path to replace legacy equipment, develop new features, and extend the market. Telephony features can be reproduced in the new products, and new features can be added, because matching some legacy technologies from the PSTN, such as E911 services, wiretapping, and caller ID necessary for broad acceptance. The voice network is growing, but circuit switching remains expensive, whereas packet switching is relatively cheap by comparison. Gateways and softswitches provide a critical link in the evolutionary path toward a single network infrastructure of converged services.

Premise-based Telephony

Before we really talk about premise-based IP telephony, we must review some issues involved in getting service to the premise to being with. The local loop presents a set of challenges that vary from location to location. While the local loop is rarely an issue for large corporations at the corporate site, it represents a major hurdle in setting up work environments for telecommuters. Because this same issue also applies to many smaller businesses, and to the SOHO sector, we'll explore it here.

We looked at several different approaches to providing QoS or assurances for the delivery of traffic. One area reviewed was bandwidth and the idea of just providing what is often referred to as gigabandwidth to solve QoS problems. Although bandwidth alone cannot provide quality and doesn't address the implementation of a prioritization scheme for different traffic types, there is a place bandwidth is sorely needed, and that's the local loop.

As shown in Figure 8.1, there are areas where bandwidth is just not a problem. Within the backbone of the Internet, bandwidth is plentiful. Providers always want more, but with the implementation of *dense wave division multiplexing* (DWDM) and expanded SONET networks, there is no shortage of available bandwidth in the core backbone.

At the premise, whether it be a corporate office or a residence, bandwidth and throughput are not issues. LANs are widely deployed: Ethernet technologies are commonly 100 Mbps at the low end of the range— such as in home networks, and Gigabit Ethernet finds increased use in corporate offices. Many computers even come equipped with LAN adapters capable of supporting this high-speed network. *WiFi*, or 802.11B wireless LANs, have become incredibly popular with residential and SOHO users because there's no need to tear up walls and install cabling. At 11 Mbps, this technology provides plenty of bandwidth on the premise.

The one place there *is* a problem is in the local loop. This isn't an issue for large corporate data centers and offices, because these facilities are often placed in business parks along major fiber routes. Large businesses typically purchase larger-capacity connections, and T-1 or greater speeds are reasonably common. These services are often specially designed and built to customer requirements, and available in part because of the higher revenue they bring to the provider.

Large businesses are often heavily involved in alternative commuting programs. Car-pooling used to be the common alternate commute, but today *telecommuting*, or work-at-home solutions are increasingly popular. The telecommuting concept is simple: business staff can provide all required services while working from their respective homes, without the need to travel to a central company location. A person's home provides one-stop service for both living and professional activities.

Aside from the alternative commuting viewpoint, there are other advantages to work-at-home programs. Studies indicate that employee retention is higher, production increases, and perhaps most important from a financial standpoint, there is no brick-and-mortar capital investment in buildings and real estate when there's a need to expand the workforce.

Much has been written about knowledge work. As our society has evolved from agricultural to industrial and now to informational, this type of work (which relies on personal computers, telecommunications, and other technologies to produce value and can be found in every business sector from factory manufacturing to finance to health care) becomes increasingly important.

Telecommuting is a fundamental component of many businesses, but even beyond that, businesses contract with other resources, which are often cottage industries. Everything from graphic artwork, to medical consultation, to application development is outsourced as companies manage their budgets closely. Integrating this remote work force into the business can be greatly enhanced with IP telephony solutions, which make the remote work appear to be on site at the corporate network.

Figure 8.1
Bandwidth isn't a
problem everywhere.

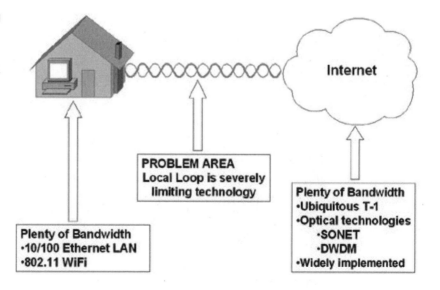

The problem with the local loop is that the bandwidth required may not be available. In many cases, it can't support the necessary tasks to perform the job before adding IP. Think about how long it takes to load a complex Web page. Remember, this local loop is a copper pair that has been optimized to support 64 kbps of throughput for *voice* traffic. It isn't only optimized for that, but given the technologies, it's also limited to that. And some of the technologies that aided in optimization are now impediments to delivering any sort of high-speed service over the loop.

This access issue has created problems for many people, but it's important to understand the root of the problem. The PSTN was designed to support voice telephone calls. Over its history, it has been continually optimized for that service. Engineers took painstaking efforts to extend the reach of the local loop and optimize every facet to minimize the effects of any impairment on the line. The overall network has been built to support the typical telephone call, something that lasts from three to four minutes on average. This call holding time is a factor in the design of the network and governs how many connections and how much equipment are activated in COs.

Telephone traffic engineers have a new problem now, and it's the Internet. Most people still don't have high-speed connections, so the most common Internet connection is via a modem using a telephone line. Now consider the average duration of a telephone call placed to the Internet. Studies indicate durations ranging from 20 minutes to 4 hours, but very few people would argue that it's anything similar to the voice telephone calling patterns.

Our own consumer spending behavior further exacerbates this problem. As the telephone network grew, it became part of our daily lives. We motivated the telephone industry to introduce flat-rate local calling because we wanted to be able to make unlimited telephone calls in our local area for a fixed monthly cost. Flat-rate billing also permeated the Internet segment. Today, something akin to the "$19.95 all-you-can-eat Internet buffet" exists today, as dial-up Internet accounts provide unlimited usage and access for a fixed monthly fee.

We don't often consider the impact this has on the network, but it's easy to step through. Here we have "free" (fixed cost, flat rate) access to the Internet. It isn't really free, but for reasonable costs, we have unlimited access. Many consumers add a telephone line just for the computer connection to the Internet. Given the pricing structure that our behavior motivated, what incentive is there for a user to log off the Internet at all? If I pay for unlimited local calls, and unlimited Internet access, why can't I just dial up and stay connected for hours at a time? Days? Weeks?

Internet access over the telephone network skews call durations, which invalidates much of the traffic engineering work that's been accomplished to optimize the PSTN for voice traffic. Furthermore, it's naïve to expect LECs to add capacity to support these long call durations since they're flat-rate lines billed like any other. The telco can't even know which lines are used for what purpose. Traffic engineering in the PSTN has become a nightmare of gigantic proportions.

Since delivering IP to the premise is a crucial aspect of IP telephony at the premise, we'll look briefly at a few of the more popular technologies used to deliver high-speed service today, particularly for smaller businesses and the SOHO environment, since this group is unlikely to implement full T-1 service in most cases. There are four solutions actively deployed today: DSL, cable modems, satellite service, and wireless Internet. Each has merits, but each has drawbacks as well. It is likely that one of these technologies will eventually become the dominant choice in the market.

Digital Subscriber Line

Digital subscriber line (DSL) technologies are one of the most visible, yet most confusing services offered today. In many instances, DSL uses an analog link to deliver digital information. DSL has the distinct advantage of being able to share the twisted pair local loop with telephone service, which means both can operate at the same time. DSL forms what is referred to as an "always-on" connection for the data link to the provider. One benefit of DSL service is that it eliminates the problem of long holding times for Internet calls by removing the data portion of the traffic from the PSTN right at the local exchange CO. The ILECs prefer DSL because it extends the usable lifetime of the copper local loop and aids in moving data traffic off the circuit-switched network.

Because DSL uses the twisted pair local loop to deliver services, it is inherently vulnerable to the physical impairments associated with that medium. Because it must peacefully co-exist with plain old telephone service (POTS), it must also comply with unique requirements. Several impediments to DSL deployment must be resolved in order to achieve success in the DSL market.

Distance is a crucial factor. Most DSL services deployed today require the subscriber to be no further than 18,000 feet from the local central office. This distance limitation affects many people in rural and sparsely populated areas of the United States. In DSL services, a simple rule of thumb is "the farther you are from the central office, the slower the connection will be."

Load coils were used in the local loop to extend the reach of voice service over long loops, usually over 15,000 to 18,000 feet. Although these coils improve the delivery of voice by tuning attenuation characteristics, they actually filter the frequencies used to deliver DSL. This means that for a provider to implement DSL on a cable, load coils must be removed. This is a labor-intensive and expensive process.

Splices are an issue in the local loop. Because the cable used generally comes in 500-foot spools, an average local loop for a subscriber 11,000 feet from the central office has 22 separate splices. While splices are not always a problem, they flex in the wind on aerial cables and are a source of potential water entry. Splices can also allow ingress noise to enter the system, thus causing further problems.

Remote terminals, known as digital loop carriers (DLCs), were widely deployed during the growth of the PSTN. These remote terminals have been used for "pair gain" systems, allowing more voice lines than there are copper pairs. They can also extend the reach of the central office by placing a remote module out in residential areas that are farther away from the central office. As shown in Figure 8.2, a remote terminal can use four pairs of wires to deliver the equivalent of 96 telephone lines to end users. A fifth pair (not shown) is used for management and backup. While the technology works wonders for delivering service at greater

Figure 8.2
Remote terminals extend the reach of the central office.

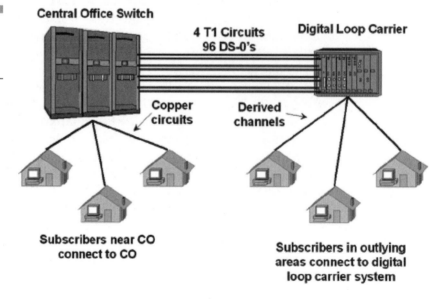

distances and for economical use of copper loop facilities, the voice circuits provided to end users are fundamentally "derived channels" or DS-0 connections from a T-1 circuit. A DS-0 is a 64-kbps channel that cannot carry higher speeds. In newer systems, the remote module is often connected via a fiber optic link, thus providing a migration path to newer technology advances, but this potential often remains untapped for now.

Digitization of the PSTN really focused on the core of the network trunks and switches, leaving the local loop to its analog legacy. The end result is that deployment of DSL has been slower than anticipated due to cost and labor efforts in making the local loop meet the necessary parameters.

DSL is available with a variety of different line coding schemes and delivery approaches. The most widely accepted model today is some variation of *asymmetric digital subscriber line* (ADSL), which supports line sharing with standard telephone service. The asymmetric aspect of this type of DSL provides for different upstream and downstream delivery speeds. For example, a downstream connection of 640 to 768 kbps and an upstream of 64 to 90 kbps is commonly available. Because most Internet users' traffic is directed downstream, to be delivered to the subscriber, this feature makes good sense for users. Symmetric options are available for users running servers who require the same speed in each direction. ADSL can deliver speeds nearly 8 Mbps in an ideal setting, but a "consumer DSL" variation more commonly used offers lower connection speeds coupled with a pricing strategy that allows several choices.

Figure 8.3 shows how DSL might be connected in the CO and how actual traffic flows to either the Internet or the PSTN. In the CO, the local loop terminates on a wiring frame. This pair is cross-connected to a digital subscriber line access line (DSLAM). In the DSLAM, a splitter separates voice and data traffic. Because the two services use different frequency ranges between 0 Hz and about 1.1 MHz, this separation of the two signals can be accomplished via a filtering arrangement. Voice traffic is directed to the local exchange CO switch and then onto the PSTN. Data traffic is directed to a router, then passed onto the Internet. For the local telephone company, this approach can be very straightforward. Their goal is to shift the data off the telephone network as quickly as possible. Remember, voice and data have completely different transmission characteristics, and Internet data doesn't need to consume the available resources in the PSTN.

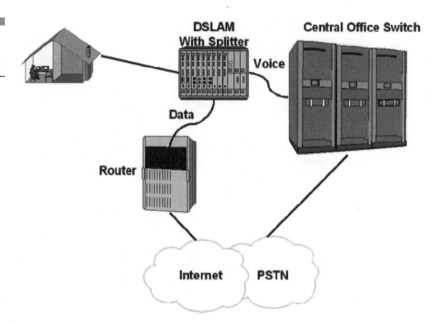

Figure 8.3
The DSL connection
in the central office.

One of the advantages of DSL is that it can easily be "ISP agnostic." The DSL service is provided over the local loop, but at the connection point of the loop to the DSLAM and splitter arrangement, the data stream can be directed to any ISP router that has an arrangement for that traffic. This means a DSL customer doesn't necessarily have to use the ILEC as their Internet provider. Consumers have a choice in many cases.

Other issues with DSL create problems in delivering the service to consumers. Many ILECs have argued that unbundling of the local loop to allow competitor access is unfair. While required as part of the Telecommunications Act of 1996, it has been administered differently in many states, creating problems for the ILECs. Many CLECs would argue that the ILECs have taken unfair advantage of their position as incumbent providers and slowed DSL deployment. Whatever the issues, DSL has been far slower to deploy than expected, and it has not met with the overwhelming success anticipated. These issues continue to plague implementations and impede service delivery.

Cable Modems

The cable modem approach is one that does not use the telco local loop, so it doesn't face the same constraints. These devices use the coax provided for cable television to deliver packetized IP. This has provided the cable TV industry with a completely new, and potentially profitable market—the Internet.

While DSL technologies are often limited to 640 kbps of bandwidth, cable modems can offer a full 10 Mbps of throughput, commonly through Ethernet technology coupled with the coaxial infrastructure. The flip side of this speed increase is that Ethernet technology brings other problems to bear. The network is a contention-based network, meaning that users must contend for bandwidth. Data can experience collisions and be discarded. Although cable modems can support 10 Mbps throughput, the more users there are, the more congested the network becomes. This leads to performance degradation and slowed response times. In early deployments, subscribers found the speeds fantastic, but as the subscriber base grew and the cable modem providers began to manage network resources more closely, many areas experience speeds far less than a full 10 Mbps.

Legacy cable TV networks were designed using what is called a *random branching bus* topology, as shown in Figure 8.4. This topology allowed for the growth of the network as new neighborhoods were added

Figure 8.4
Legacy cable TV
network topology.

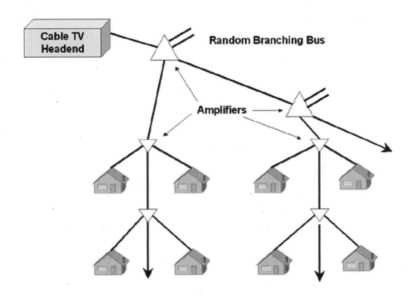

to the cable television network and extended the reach of the network. As you can see, this network provided for growth as demand increased. While this technique provided for the delivery of cable television service, the growth of the network often dictated that the network pass some 20,000 homes before reaching the end of the link. Additionally, television is inherently a broadcast technology with all the traffic flowing downstream to subscribers. There was no need for a usable upstream channel in the television network. All the amplifiers were pointed in the downstream direction to provide better television reception.

The cable TV network was built specifically to deliver a particular type of traffic, in this case, broadcast television in the downstream direction. Clearly, the network was not designed with the idea of carrying two-way Internet traffic. As a result, cable providers have undertaken some massive redesign projects to integrate fiber optic solutions into existing topologies. This *hybrid fiber-coax* (HFC) approach has allowed tremendous improvements in Internet delivery in many cases, but is still not ubiquitous in the network. Fiber is generally added to the backbone, not to the subscriber's household cabling, and fiber cannot carry electrical signals, only data signals. The amplifiers and other network equipment in the cable television network are often line-powered, thus necessitating copper connectivity.

Cable modem technology allows for about 100 downstream channels, and in most cases only about 10 are used. This network obviously brings the potential for expansion and delivery of Internet services, but because this was not the original purpose or design, hurdles must be overcome in order to provide all the desired services at acceptable levels. Cable providers are actively working toward this end.

Because this alternate local loop technology is not part of the PSTN, it isn't subject to the regulations of the telecommunications industry. There is no equal and open access requirement for the cable network; if a provider can deliver IP telephony over the cable connection, no access charges need be paid to the ILEC for terminating the call. Because these access charges consume up to 35 percent of the revenue received for processing telephone calls, this tremendous cost savings has been very attractive to many companies, perhaps most notably AT&T.

To the consumer, this means that there is no longer a choice in ISP. The cable television provider is the only ISP available in many cases, despite lawsuits attempting to gain open access (AOL is certainly working hard to reach these customers). Time will certainly force a more open arrangement than we have today, but at present choices are very limited.

If the cable-based providers achieve measurable success in delivering IP telephony using this model, the telecommunications industry will suffer financially and there will certainly be lawsuits and regulation changes to attempt to provide everyone with a fair playing field. Whether these changes will be for the good of consumers and service availability or for the financial good of the telecommunications industry remains to be seen.

Satellite Internet

Another approach to Internet delivery brings a completely different local loop into the game. The satellite delivery of Internet service has some distinct advantages and disadvantages over the wired technologies in use.

This model can be quite convoluted, as shown in Figure 8.5. In initial implementations, the subscriber required both a satellite dish and a telephone line for full connectivity. The subscriber must initiate a telephone call, via modem, which connects to the provider. This connection is used for the upstream traffic, carrying anything the user transmits to the Internet. A request for a Web page is sent via the telephone line to the satellite provider's point-of-presence (POP). This request is then routed to the Internet and passed on to the destination Web server. The

Figure 8.5
Traditional satellite
Internet topology.

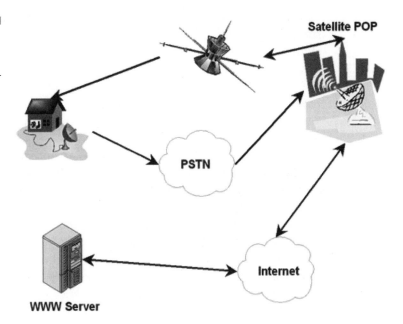

reply is routed back to the satellite POP, and uplinked to the satellite, then downlinked to the subscriber's satellite dish antenna. Most satellite providers deploy two-way satellite links that eliminate the requirement for the telephone line, but the "dial return" arrangement shown is quite common among satellite subscribers.

Satellite service is quite asymmetric in nature. The upstream connection is limited to modem speeds, whereas downstream connections vary based on network load, individual satellite load, and atmospheric conditions. Typical speeds range around 400 kbps, with slower speeds common, and occasional speeds of 700 kbps noted by many users. Much higher speeds are anticipated as this technology matures, but speed comes at a price. The consumer service described here costs somewhat less that $100 per month. Higher speeds will cost more, with projections ranging from hundreds to thousands of dollars monthly.

Satellite service has some very distinct differences from the wired technologies we've discussed. Because there is no physical infrastructure, cables never get cut by contractors, but atmospheric conditions can and do impact service. An afternoon thunderstorm frequently knocks the link out for hours.

Although the service provides a high-speed connection that works well for Web surfing or downloading files, there is inherent transmission delay. Satellites circle the globe in geostationary earth orbits some 23,000 miles in the sky. Signals transmitted this distance can't be instantaneous, and the delay will impact some applications. Typical Internet delays in a dial-up connection range from 70 to 150 milliseconds for *ping* to send an echo request and get an echo reply back. (Ping is a simple reachability test that can be run from any network node.) In the consumer satellite service offerings, ping response times generally range from 700 to 850 milliseconds. Given what we know about voice transmission, this delay is far too great to support IP telephony or any real-time, two-way, interactive application in use today.

From the provider's perspective, a big advantage to satellite technology is the coverage area or footprint. Once the network is deployed, (i.e., the satellites are in orbit and functioning), the potential customer base totals roughly six billion people (the world population). Planetary coverage gives a provider a pretty substantial market to draw from. Even a 1 percent penetration rate in the market yields a subscriber base of 60 million users—almost double that of the largest Internet provider, America Online. While not everyone on earth uses the Internet, the marketing presence achieved by delivering service to every home on earth certainly has some business appeal.

Fixed Wireless Internet Technologies

Fixed wireless, like satellite, requires the user to be in a fixed location. While mobility and ubiquitous access to the Internet are very popular goals, they are not an area we will explore in this book. Third-generation (3G) wireless technologies are deployed in some parts of the world, and tested in others, but they remain on the horizon. Several fixed wireless solutions are in various trial and implementation stages around the United States.

- *Multichannel, multipoint distribution systems* (MMDSs) operate in the 2.5-GHz range. MMDS began as a 32-channel video service. Sprint and WorldCom have both conducted trials of this technology. SpeedChoice is the largest active provider at present.
- *Local multipoint distribution system* (LMDS) also began as a video service and was marketed at one point as "cellular cable TV." This technology operates between 20 and 40 GHz and has been a topic of study in the IEEE Broadband Wireless Access group. Licenses for much of the frequency spectrum attracted considerable attention in 1998, with some very high bidding in the process. WinStar was one of the notable providers using this approach. While it has struggled with a workable business model, as of May 2002, its restructuring plan began showing some successes.
- *Unguided light wave* is a very successful new technology. This approach is often referred to as *free space optics* and can provide a 100 Mbps connection within line of sight. It's been described as an optical counterpart to line-of-sight microwave systems. Whereas microwave radio systems require FCC licensing, these optical systems do not. The most notable provider of this service at present is Terabeam (www.terabeam.com) in the Seattle area. Terabeam has systems deployed in several U.S. cities.
- *Wireless Fidelity (WiFi)* or *802.11*, perhaps the most notable of the fixed wireless technologies at the time of this writing, was originally introduced as a wireless LAN (WLAN), technology but with some creative implementation, it has become an Internet access technology that is gaining momentous support around the world. The Wireless Internet Service Providers Association (www.wispa.org) was founded in February 2001. An estimated 3,000 wireless providers in the United States each try to provide service in the 2.4-GHz range of the spectrum, and the number grows daily.

WiFi

Barrier to entry for WiFi service providers is relatively low. For the one-time expense of a few thousand dollars and the monthly recurring charges of a T-1 connection fee to the Internet, an individual can quickly deliver service to a town center, neighborhood, or business park. This wireless LAN technology is very similar to Ethernet, in which users contend for the network capacity, but live implementations have shown that a 384-kbps fractional T-1 to the Internet can support about 20 users. Based on usage patterns, users get about 200 Kbps throughput in this configuration.

The 802.11B LAN technology is an 11-Mbps technology; however, since it is wireless, it is distance sensitive in some regard. The farther a subscriber is from the antenna, called a *wireless access point* (WAP), the weaker the signal and slower the speed. To get the 11-Mbps speed, the user needs to be within about 300 feet of the WAP; however, speeds of 1 Mbps can provide a much greater reach.

WiFi is such a disruptive technology that it has become one of the hottest segments of Internet deployment efforts. Access in the last mile radically changes, and the wired local loop or any wired topology may quickly become irrelevant. Given the range of local loop issues, that last mile connection to the subscriber has often become the "lost mile," lost in the sense that you just can't get there from here. Enter WiFi and a technology that crosses the abyss of users who don't live where cable modems are available and whose local loops cannot qualify for DSL service.

802.11B is not without a new set of problems. First and foremost, it operates in unlicensed spectrum at 2.4 GHz. Unlicensed spectrum can be used for a variety of services, and 2.4-GHz cordless telephones are also popular at present. Does this mean you could implement WiFi and your neighbor can implement a cordless phone system than could interfere with it? Yes, absolutely. And there is no real recourse available because of the unlicensed spectrum. That doesn't mean that problems will or must occur, and they can certainly be resolved in most cases. But as the usage density in this frequency range increases, so will the number of conflicting devices.

Another area of concern is the power constraints and requirements. While reputable providers and organizations don't encourage doing anything that might be questionable, a number of operators quickly boost power and use equipment that isn't really industry standard in terms of quality. To gain a clear picture of this, you might try visiting an Internet search engine and search for the terms "WiFi," "802.11," and "Pringles" in combination. You're likely to find instructions for building a high-gain antenna using Pringles cans. Figure 8.6 demonstrates how wireless users might access the Internet, but most designers anticipate servers remaining at fixed locations, connected via wired media for the foreseeable future.

Figure 8.6
WiFi (802.11b)
wireless Internet
topology.

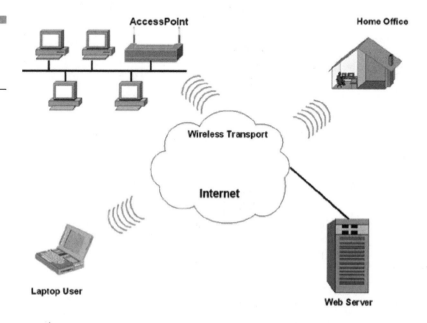

Still, the ability to turn on your battery-powered laptop computer sitting on your patio or at the beach, or anywhere at all, is an almost mystical experience. This technology has become so popular that activists are using it to broadcast "free" Internet access to anyone within range. ISPs are warning customers that doing so violates contractual agreements. In London recently, a movement referred to as warchalking sprang from grassroots support. The idea there, which has quickly spread to the United States, is to mark with chalk symbols the sides of buildings that provide wireless access to the Internet. Initially, three symbols were used to identify open nodes, closed nodes, and *wireless equivalent privacy* (WEP) encrypted nodes. This, coupled with the *service set identifier*, provides enough information for the technically savvy user to connect quickly to someone else's Internet connection. The idea of chalk marks comes from earlier times, when hoboes riding the rails used symbols to share information. An icon of a cat was always used to indicate a "kind-hearted woman lives here," which often meant a vagabond could get a free meal.

Clearly, like IP telephony, wireless Internet has tremendous grassroots support. Disruptive technologies are those outside the mainstream, typically low in cost, and used in ways not originally envisioned. IP telephony and WiFi Internet are two of the most disruptive forces at play in the Internet and telecommunications industry today.

If We Build a High-Speed Network, Will They Come?

Considerable attention is focused on the availability of high-speed connectivity today. Note, we said *connectivity*, not *services*. Many areas of the country lag far behind in deployment of DSL, cable modems, and other high-speed solutions. Yet, in areas where these solutions are widely available, market penetration rates remain dismally low, at around 12 percent in general. This penetration rate seems to indicate that if we build it, "they" may not come after all. There must be a reason to come. There must be demand. And demand can only come from applications and services. IP telephony is one example of a service that can drive increased demand.

Building demand is a major issue. Many states have undertaken telecom aggregation strategies to identify the revenue stream and pent-up demand for high-speed connections. Unfortunately, in most cases aggregation aids in identifying only current demand. It may help uncover some anticipated future demand, but it cannot create demand where none exists. Only education about services and applications can create demand. IP telephony is one such service, along with many others.

Regulatory issues and legislation changes that can have a positive impact must be addressed on a continual, sustained basis. The current state of the Telecommunications Act of 1996 provides a disincentive to ILECs to provide IP telephony. The Tauzin-Dingell Bill (HR1542) passed in the House of Representative recently, but is unlikely to pass the Senate. Both the federal government and individual states must create legislation that establishes fair and level competition among providers, and does not unduly penalize one group over another.

Last-mile and middle-mile issues must be addressed concurrently to achieve success. Given the low demand we're finding, achieving a middle-mile solution is a "build it and they will come" approach. Given that the penetration rate for high-speed access, where available, is only 12 percent, we know that if we build it, nobody will come, and more providers will suffer and fail. Providers must build business cases that support a return on investment from building new infrastructure. If the profit margin doesn't justify the investment, there is no incentive to build a high-speed network. On the other hand, if wireless initiatives, such as those we discussed earlier, truly loom in the market, providers will be forced to take some serious action to protect their embedded customer base.

All of these issues surrounding the delivery of high-speed connectivity affect businesses considering IP telephony implementations. Small companies often can't afford T-1 connections and rely on DSL or cable modem availability for the best possible service. Many small companies still use dial-up as their primary means to access the Internet. Larger companies may not have access issues at their corporate locations, but as the technology continues to evolve, more and more people wish to work from home. Telecommuting programs are often major incentives for large companies when recruiting efforts are necessary. All companies search for the best and brightest people to fill open positions. If those with the talent to best do the job live in another area, telecommuting may be the only tool a company has to attract the best work force. These issues affect every company in one way or another, because without a solution to the local loop dilemma, no adequate bandwidth is available to allow remote workers to participate as full members of the corporate network. The term *virtual office* has been widely used, but without the critical link, there is virtually no service to the remote site.

Once service is provided, or acceptable service levels are met, we can begin to explore how IP telephony might fit on the local premise.

Work-at-Home Staff— Telecommuting

There are many reasons why a company might want to support telecommuting programs, and just as many reasons why people might want to work at home. Many people telecommute on a casual basis—working from home on snow days in winter, or when a sick child needs attention, but more and more mainstream companies are adding to their ranks telecommuters who work from home all the time.

The first distinguishing characteristic of the telecommuter trend is the lack of need for mobility. A person's home provides one-stop service for living and for professional activities. The second important characteristic is the career involved, including training and supervision, along with salary advances. Third, there is the ability to keep the family together as a unit; in other words, to be able to watch over an ill person confined to the home, the potential to avoid the need for day care facilities, and more.

From the point of view of the organization employing telecommuters, three primary advantages exist. Studies indicate that employee retention is higher, production increases, and, perhaps most important from a financial standpoint, no "brick and mortar" investment is needed in order to add to the workforce.

A great deal has been written in the last two years about knowledge work. As our society continues to evolve from the Industrial Age through the Information Age on into the Knowledge Age, this type of work becomes a dominant facet of the economy. Information, which is transformed into knowledge, becomes intellectual capital of very high value.

On the other hand, not much has been written about knowledge workers themselves, specifically those who have managed to create new work constructions and adaptations that replace the commuting-based structure that has been the default for the last century or more. These are not employees who work at home only when the plumbing breaks or a budget meeting can't be postponed. These are the people who have made location increasingly and regularly irrelevant to the successful completion of their day-to-day work. They are knowledge workers not in theory, but in practice.

IP telephony holds tremendous promise for those workers who have high-speed services available. Many companies have used ISDN lines for telecommuters in the past, but ISDN never truly provided a total integrated solution for voice and data services. IP telephony can provide true and complete convergence at the desktop to a single device, the computer.

Implementing IP telephony in the business workplace, and extending the service to the work-at-home staff allows for integration beyond what most companies provide today. The employee's office telephone number can ring at their home office on their workstation. IP telephony can provide integration that allows employees to flag their availability or absence, and an integration of voice and data services that was never truly reached using ISDN.

Home workers can log on in the morning, and use standard office applications just as they might in the office. Collaborative software suites like Microsoft Outlook or Lotus Notes are widely deployed in offices, with shared calendars and folders. (These applications increase productivity and collaborative efforts, but are poor performers over dial-up lines.) Couple this capability with a software-defined telephone extension in the IP telephony system, and colleagues in the office can reach these workers by dialing simple four-digit extensions. They become linked directly to the office just as if they were sitting on site in a cubicle. Even added IP telephony variations like Microsoft Netmeeting and voice conversation tools in MSN Messenger, both quite standards-compliant programs, become incredibly useful collaboration tools.

IP telephony for the telecommuter may well be the leading edge of implementing real-world convergence—network convergence that integrates the voice and data networks, desktop convergence that consolidates the computer and telephone, and the convergence of business needs with personal needs for this highly motivated workforce.

Replacing the PBX in the Office

For years, businesses having been searching for ways to consolidate support services and reduce cost. The telephone system, or PBX, has been an enormous obstacle, sometimes costing hundreds of thousands of dollars over the lifetime of the system. Administering LANs and connections to the Internet has become simpler as technologies improved, but the training and education required is far more important—and costly—than improved technology. College graduates often take courses in IP networking, implement small networks, and even develop Web applications or new protocols.

Telephone systems are often leagues apart from personal computer technologies and LANs. Many telephone systems are network connected only for administration, which is still performed via a VT-100 terminal emulation. And many have some form of Web interface for GUI-style management. Although the GUI lends itself nicely to "point-and-click" usage, the ability to perform tasks efficiently and properly still requires an understanding of telephony and how PBXs work. Some of these systems may be testimony to the cold, hard reality of voice and data networking—more skills are required to be successful than just point and click. These systems are often unfamiliar and cumbersome to a data specialist who may be quite comfortable working with servers and routers.

The benefits of consolidating this infrastructure onto a single platform are tremendous:

- A single wiring plant can replace the two sets of physical wires needed for connectivity.
- The PBX itself can be very expensive to purchase. Even smaller key systems are expensive and awkward for many businesses to deal with. Many of these systems require proprietary digital telephones from the same vendor that provided the PBX.
- Administration skills can be consolidated onto a single platform or technology set. A LAN technician may find a Windows NT or Linux-based system far easier to work with than another proprietary operating environment.
- Billing consolidation may be one of the largest administrative savings achieved. Auditing and processing two separate sets of bills requires extensive administrative support.

In the early '90s, some attempt at reaching this goal was undertaken by the IEEE in the form of the 802.9 standards for *isochronous Ethernet*, or *IsoEthernet*. This blending of the 10BaseT Ethernet LAN and ISDN onto a single cable was universally applauded as a great concept, but the actual implementations never gained market acceptance. The components were expensive, marketing was disjointed, and many vendors ignored the solution entirely. Perhaps one cause of this failure was the lukewarm reception ISDN services have always received in the United States.

With the onset of real convergence in networks, the idea of PBX replacement has again become popular. In its simplest incarnation, a PBX attached to the LAN really doesn't need to be anything more complex than an IP telephone gateway that supports a LAN connection on the premise side and a T-1 or some type of local trunk connection to the PSTN. The PSTN would see this LAN-based PBX as it does any other PBX.

One advantage to this approach is that a computer isn't necessary for workers to use a telephone. Many different IP telephones are available today that can plug directly into an Ethernet LAN. These devices include a codec to perform the voice sampling and encoding, a TCP/IP software stack, and the LAN adapter and drivers required to make it all work. While they look like telephones, the inner workings are different, and these devices become fully functioning nodes on the IP network, complete with IP address.

With the downward spiraling cost of computers, especially in the office environment, nearly everybody has one on their desk today. To function as a telephone, a computer has to have some software running to provide the encoding and decoding functions. This software is readily available from various sources, some free, and some as commercial products. The PC must have a sound card for audio capability, and may be equipped with speakers and a microphone, or support the connection of a headset or handset. Some implementations use a telephone connected to a serial port of the PC.

The LAN administrator can manage all the telephones as part of one consolidated addressing scheme on the IP network and, in the cases of the LAN-connected telephones, users might never even be aware that they are using this new technology.

This solution provides very simple, basic telephony. Users pick up their telephone, or click a button on their computer to place a call. The dialed telephone numbers are routed over the LAN to the gateway, which can either process them internally using IP, or direct them outward to the PSTN. This simple implementation of a PBX replacement is shown in Figure 8.7.

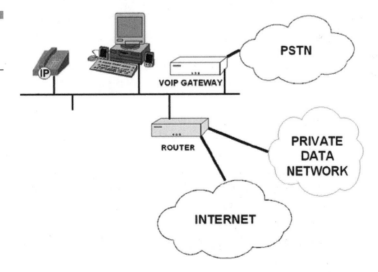

Figure 8.7
Telephony and data service on the LAN.

The strongest benefit isn't the simple consolidation onto the single wire, but all the integration of services that can follow, which include the collection of customer data, automatic call distribution functions, and combining voice and data operations into a new next-generation application for the business.

It isn't all a bed of roses, however. There's an old adage that says "you get what you pay for" that rings truer than ever when implementing IP telephony as a PBX replacement. And the key is the word *telephony*. Packetizing voice is simple, but to folks in the industry, the word *telephony* conjures up a rich set of services that go way beyond putting voice traffic inside data packets. Many of these features go well beyond the quality of voice traffic and are taken for granted to today's business telephone systems:

- *Simple features* like call waiting, call hold, conference calling, call transfer, and music on hold are overlooked in business because they're ubiquitous in most business phone systems today.
- *Conference calling* capability is generally available by hitting a flash button or tapping the switch hook to add a second party. More sophisticated systems allow adding a third or fourth party, or have conferencing bridges to allow multiparty conferences.
- *Caller ID* for inbound calls is standard for most primary rate T-1s connecting PBXs to the PSTN and allows for enhanced call processing. Outbound caller ID generally displays the main number for the PBX system.
- *Automatic route selection* (ARS) or least-cost routing is a standard feature in PBXs, allowing multiple connections to multiple toll and long distance providers. This feature enables the PBX to select outbound calls for path-based pricing and may incorporate *time-of-day* routing, so that calls are handled differently during peak calling periods.
- *Voice mail* systems digitally sample and encode messages onto a hard drive within the system. They, too, can be IP networked devices. It may be necessary to replace an older voice mail system as well as the PBX to ensure everything continues to work. Some vendors offer integrated solutions that work together seamlessly.

While all of these features (and more) are available in IP telephony solutions to replace the PBX, they may not be standard, and they may work differently from the traditional features in an older PBX. The key is to make a list of all the features expected or needed and ensure that whatever solution chosen supports these necessary features. The presence or absence of the complete range of standard telephony features such as these serves as a clear distinction between systems that carry voice inside IP packets and complete IP telephony. The ability to extend these features over the IP connection to a telecommuter or remote location brings the convergence of voice and data another step closer. The greater the level of features integration in a seamless and transparent system, the smoother any transition will be.

One area of IP-based PBXs remains a significant area of concern—E911 services. Emergency agencies in the United States rely on caller information provided by the PSTN when someone dials 911. Regulations have been expanded in recent years, with more effort to pin down the caller to as precise a location as possible.

One of the benefits of IP telephony is the ability to log into your telephone network from anywhere. With the simplicity of the LAN environment, administrative functions like additions and moves are easier to accomplish. On the down side, if users dial 911, the emergency response teams may have a very difficult time locating them to bring help. The solutions that do provide real E-911 support tend to rely heavily on POTS service from the PSTN. Perhaps that leads to one variation of a solution—Why not use the IP PBX internally for telephone calls within the office building or campus but, rather than attempt to force all traffic to the Internet, just connect the IP-PBX to the PSTN for all telephone traffic to the outside?

Local users on site can be located within the network, based on connectivity information. An IP telephony system can be designed into the network so that the hardware port that the user is plugged into is easily identified. But as companies add to the remote work force and telecommuting becomes more and more common, remote access callers and others using extended software phone services can't be located with any degree of accuracy. Unfortunately, this is an area that the IP telephony vendors seem to pay little attention to, leaving support and solution integration for emergency response in the hands of the customer.

Although there are issues that require special attention, this doesn't mean that IP telephony isn't a solid viable replacement for the PBX. It makes perfect sense in the right situation, and is disastrous in the wrong setting. Only the informed customer can determine whether or not it's really the right fit. Consider the merits, weigh the costs, and, once again, make an informed business decision. The worst mistake a company can make is for an executive who's become enamored of the technology to force implementation because of some personal affinity for something new. Good business analysis often demonstrates that the

PBX is an incredibly expensive resource that can sometimes easily be replaced, thus allowing for the consolidation of many facets of the network. The savings can be enormous and are worth the effort required to conduct a complete analysis.

Computer Telephony Integration (CTI)

Integration of computers and telephony has been a very long time coming. With the features of the advanced intelligent network (AIN) and CTI systems, we've been able to develop computerized systems that provide information based on input from telephone touch tones. This early step towards convergence was taken many years ago, and many people don't even realize the role it plays today.

Consider the places you call that use some form of CTI. Call your insurance company, and you're often asked to enter your policy number so that the representative has your information on the screen when they answer the phone. Call the IRS to check on your taxes, and you're prompted to enter your social security number. Have you used the automated banking line to see if your paycheck was posted via the electronic funds transfer (EFT)? Have you checked the status of stock investments or a home loan application? Many of us use these common CTI applications almost daily.

From the business perspective, CTI can work in the other direction as well. Businesses use CTI applications to generate calls automatically for fund-raisers, based on a database of members or former contributors. Telemarketing and telesales companies use them to blast out those dinnertime phone calls we all love so much. Political groups use them in calling campaigns to win votes or reach masses of constituents.

Many of these CTI systems also incorporate *interactive voice response* (IVR) systems to enable even greater collection of information when the person on one end speaks into the phone; voice recognition software allows that data to be interpreted for input to the system. These IVR systems are far more powerful today than when they were introduced several years ago.

Although computer telephony integration has been around for some time, the applications that drive the linkage between the telephone systems and the databases or applications at the business have often been custom written for very specific purposes. The problem is that these applications must be written by specialists to perform specific tasks. They sometimes require a specific model of telephone system, a clearly defined computing platform, or both. Administering the system may be GUI driven, but customizing the interactions between the voice and data systems is often akin to alchemy—an art that only the most gifted of wizards dares undertake. The end result is that the solutions are often too expensive for many businesses that could have benefited from the technology.

As shown in Figure 8.8, a new CTI model can be implemented today that provides a level of integration that was never possible in the traditional environment. The IP PBX is connected to the LAN for handling IP telephony calls that come in via the Internet, and to the PSTN for traditional telephone calls. The CTI server can interpret dialed digits from telephone users or IP information from Internet customers. Either remote user can gain access to the customer service representative, and the service rep can have complete customer information regardless of the source of the call.

Figure 8.8

Computer Telephony
Integration (CTI) built
with IP.

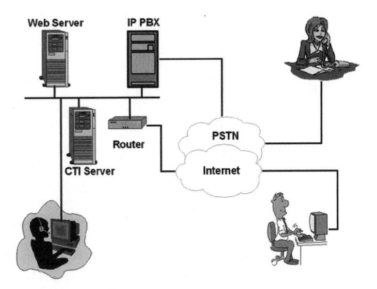

The level of CTI possible in the IP telephony setting takes the best advantages of each technology and bundles them to provide potential services previously not feasible. We'll discuss some of these new possibilities in the last chapter.

Why Not Implement All Three?

Many telephony solutions are in use in business environments today. We've just reviewed three, but in reality, those three can be interlinked in such a way that no lines of distinction exist. In Figure 8.8, the customer service center representative was shown in the office connected to the LAN. Is there a reason to be? This job could just as easily be performed by a team of remote workers who might not participate in the "9-to-5" office work force. Stay-at-home mothers, retirees seeking part-time work, and folks who might not be able to drive suddenly have the potential to more easily become active participants in the day-to-day activities of the business.

The IP PBX might be connected to an IP network like the Internet. It might be connected to the PSTN. It might be connected to both. It's not unreasonable to speculate that the Web server and CTI server could run in the same server, with the IP PBX and voice mail system in another.

How users connect to the network is irrelevant. IP is unaware of the delivery mechanism used because it truly does not matter. Connectivity can be provided via any network mechanism, but the most crucial piece of the puzzle is making that connection. DSL, cable modem, and fixed wireless solutions all hold promise, whereas dial-up solutions are quickly being relegated to some "lower class" of service for converged networks. What matters is that users can connect with the service parameters required for the applications in question, and those applications vary.

The list of premise solutions available is as endless as human creativity allows. Vendors constantly adapt and adjust to the market, and every variation a business might require can be designed and built. And they all work! The key, reinforced time and again in this book, is that a successful implementation isn't just one that functions. It's one that improves the business in ways that can be measured and monitored. The way to achieve that is to build a business case that demonstrates success long before any selection of vendor equipment is made. Do this, and success is assured before you change anything in the network.

- *Account codes* are used for managing expenses and these are often used to bill telecom expenses back to individual departments or work groups.
- *Anonymous call rejection* refuses calls that do not deliver caller ID information.
- *Automatic callback* is used when a busy signal is received. The caller can enter a code that automatically retries the number, and then rings the caller's phone when the line is free and the call can be connected.
- *Automatic ringdown or hotline* allows a station to be programmed to dial a number automatically when the handset is lifted. This feature is often used for lobby phones to ensure that these stations, while accessible to anyone, aren't used to incur toll or long distance charges. Emergency phones in elevators are another example of how this feature might be used.
- *Barge in*—allows a user to join a call in progress on some other station, creating a three-way conference.
- *Call blocking* can be programmed with a list of specific numbers to block.
- *Call forwarding (busy, don't answer)* comes in an array of choices. Calls might be forwarded to one number if there is no answer, or another if the line is busy. Calls may ring then forward to a voice mail system. Call forwarding can also be user programmable so that the user can direct where calls are forwarded to when they arrive. *Selective call forwarding* can be used like call blocking to forward calls received from a preprogrammed list of telephone numbers.
- *Call hold* allows users to perform other tasks while not releasing the telephone call.
- *Call park* provides a mechanism for users to place a call on hold at one station, then move to another phone and pick the call back up by dialing in a code.
- *Call pickup* can be used to allow employees to pick up a call when they hear a telephone ringing. In many offices, this allows people working after hours to pick up incoming calls to the main number at their desk after an operator or receptionist has left for the day. Frequently, any station in a call pickup group can pick up calls to any other station in the group.
- *Call or station restrictions* can be used to prevent some phones from dialing outside the building, while preventing others from making toll or long distance calls. Phones can also be blocked from receiving calls. These features are sometimes used to control employees' use of the telephone for personal calls.

- *Call return* lets the user enter a code to call the last party that called the user.
- *Call transfer* simply provides a means to transfer calls from one station to another.
- *Call waiting* provides a signal, usually a beep tone, to indicate an incoming call while a telephone call is in progress. The called party can then typically place the first call on hold, answer the second, and toggle between calls as necessary.
- *Caller ID* lets the called party identify the name and phone number of the caller prior to picking up the telephone.
- *Consult hold* allows the user to place a call on hold by either depressing the switch-hook or pressing a "flash" button. The user can then make a call to another party and be reconnected to the original call after consultation.
- *Distinctive ringing* uses a special ringing pattern to flag the called party as to where a call might be coming from. Typically this might be used for differentiating outside calls from internal calls.
- *Intercom* service allows for internal abbreviated dialing, used by dialing the four-digit extension of the station.
- *Hunt groups* allow calls to ring sequentially from station to station when lines are busy. This allows numbers to be grouped together in a pool, usually a common work group, so that incoming calls aren't sent to voice mail. This arrangement is frequently used in service organizations.
- *Last number redial* lets the user redial the last number called.
- *Message waiting* comes in two standard forms and is designed to work in conjunction with a voice mail system. Audible message waiting sends some notification to the user when the receiver is lifted. This audible signal is usually what is referred to as a *stutter dial tone*. Stutter dial tone is most commonly used with simple analog telephones that do not have any additional features. More advanced telephone sets are often equipped with a message-waiting lamp that the system lights when a message has been left in voice mail. Centrex services provide a standard interface so that third-party voice mail systems can turn the message waiting indicators on or off as necessary.
- *Music on hold* can also be used to provide product or service information to callers who are waiting.
- *Speed dialing* provides a user-programmable list of numbers that can be used to dial frequently called telephone numbers via a speed calling code rather than dialing the whole number.

- *Station message detail recording* (SMDR) provides a tool for producing call detail records for each station. This gives managers information to control abuse, monitor cost accounting, and analyze calling patterns that might be useful in reducing costs.
- *Three-way conferencing* lets the user add another party on to an existing conversation.
- *Toll restriction* is used to block a station from placing toll or long distance calls.
- *700/900 blocking* is used to block calls to fee-based services.

When using ISDN Centrex services, other features may be available. These include multiple call appearances, allowing a telephone number to ring on several stations, and secondary telephone numbers that might be used for private telephone lines.

Centrex Evolves to IP Centrex

We've already reviewed how voice is packetized and IP telephony works. The next step is how to integrate IP telephony with Centrex in the telco CO. In general terms, IP Centrex means that the customer uses IP as the core networking technology to send calls to the network. That network might be the PSTN or an IP network like the Internet. This transmission is typically over a broadband packet connection.

One of the major benefits of this approach comes in terms of bandwidth utilization. We reviewed voice encoding schemes earlier, and they are all available in almost all IP telephony equipment today. This approach converges voice and data onto a single broadband access facility. If the encoding scheme chosen allows for it, many more simultaneous telephone calls can be active over a connection. If the bandwidth isn't being used for voice calls, it can be readily available for data transmissions. The gain is in the efficiency of capacity utilization.

Traditional Centrex requires a pair of wires for every analog station. If the phone isn't in use, the carrying capacity of that pair is unavailable. The IP Centrex solution can provide a level of service convergence that increases efficiency dramatically. Two network architecture approaches are commonly used today to provide IP Centrex services: the Class 5 switch architecture and the newer softswitch approach.

IP Centrex in the Class 5 Switch Architecture

The Class 5 switch platform makes up the core of the PSTN. Since it can easily support Centrex services, the addition of IP, provided in upgrades over the past several years, is the only requirement. What's new in this scenario is a variation on the gateways we described earlier. Here we see what's called the *network gateway* and the *customer gateway*. Both perform the basic functions of all gateways (discussed in Chapter 7), but these are optimized for particular types of service. These two gateways send signaling information using SIP or H.323. Since they interconnect to different types of network, each is generally capable of supporting a wide range of interfaces to support legacy telephone equipment.

In Figure 9.1, we see a new device called an *integrated access device* (IAD) on the customer premise. While manufacturers make considerable noise about these devices, in its simplest form, the device is an IP telephony-enabled router. The IAD can support both voice and data on a single chassis. Thus, in addition to standard router interfaces for Ethernet, Token Ring, and WAN connections, the IAD can support foreign exchange station (FXS) and foreign exchange office (FXO) ports for telephony services. The IAD may be a standalone device, or it may be a coprocessor device optimized to work integrally with a router.

Figure 9.1
IP Centrex in the
Class 5 CD Switch.

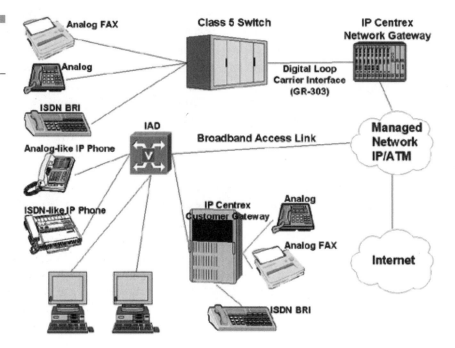

The network gateway connects much like the digital loop carrier systems described in Chapter 8, using the GR-303 protocol set for communications with the CO switch. In one direction it converts packetized telephony into circuit-switched telephony for the PSTN. In the other direction, it converts the circuit-switched information into packets for delivery to the customer gateway.

The customer gateway performs comparable tasks, but with an orientation toward the station devices it can support.

The end result is that everything from the network gateway out through the packet network at the customer premise appears as a digital loop carrier (DLC) system to the Class 5 switch in the CO. The telephones connected to the customer gateway look like any other telephones to the network. Because it appears to be a DLC, the CO switch can deliver the same feature set to IP Centrex users that it delivers to traditional Centrex users. Thus, Centrex services can be delivered to an entirely new set of subscribers without incurring the cost of an expensive CO switch upgrade or replacement.

IP Centrex in the Softswitch Architecture

In this configuration the IP Centrex provider no longer uses a Class 5 switch, replacing it with a variant of the softswitch, which performs the functions of a full switch using software. This approach is popular with competitive local exchange carriers (CLECs) that want to enter the telephony services market, but do not have an existing network comprised of large and expensive Class 5 switches. The softswitch is a telephony application running on a network server to provide call control functionality.

Whereas the Class 5 switch is directly involved in the transport and switching of the packetized voice stream, the softswitch is not. Instead, it communicates with the customer centrex gateway using one of the IP telephony protocols like SIP or H.323. It receives the call setup request from the customer's equipment directly over the data portion of the network, and then passes necessary information to the signaling gateway for translation to the PSTN using SS7. The trunking gateway performs the pack-to-circuit conversion of the actual voice conversations. Both the trunking and signaling gateways communicate with the softswitch.

This type of softswitch is becoming very popular with providers who do not have an embedded base or established network of Class 5 switches. The ILECs are also using this approach to provide IP Centrex when the CO switch can't directly support the necessary features. The softswitch approach is considerably less expensive than a full-blown CO Class 5 switch. The softswitch IP Centrex architecture is shown in Figure 9.2.

Figure 9.2
IP Centrex using a softswitch.

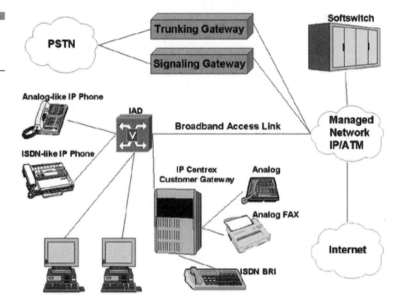

Customer Premises Equipment for IP Centrex

When implementing traditional Centrex solutions, the only equipment visible on the customer premise is the telephones. IP Centrex adds some new items that help provide the services. In the IP Centrex architecture, the customer premise still has telephones, perhaps a mix of legacy telephones and newer IP phones. In addition, we see deployment of the customer gateway and the integrated access devices (IADs). Additionally, there may be ISDN terminal adapters if the IP Centrex solution is designed to support ISDN phones.

The customer gateway provides support for existing analog and/or IDSN telephone stations. It usually has an Ethernet port for connection to the LAN, and frequently has RJ-11 analog telephone jacks to support connecting simple telephones and fax machines to the network. If analog devices are connected to these ports, the gateway digitizes and packetizes information sent from them. If ISDN devices are connected the gateway, it also converts Q.931 signaling messages into IP telephony packets. The gateway transmits the IP telephony traffic over the LAN to a high-speed access link that connects into the provider's network. If the provider is using a Class 5 switch, traffic would be routed to the network gateway, otherwise it is directed to the softswitch.

The integrated access device (IAD) is used to connect the customer gateway to a router, Ethernet hub/switch, or the connection to the WAN. This connection might be a T-1 interface or an Ethernet connection to a DSL or cable modem.

IP telephones come in a variety of styles, and they might also be referred to as *LAN phones* or *Ethernet phones*. These phones may incorporate some of the gateway functionality, in that they can often perform digitization and packetization functions. This requires a hardware design that incorporates a *voice codec* and a TCP/IP stack into the telephone. The IP phone has an Ethernet interface for connection to the LAN, using an RJ-45 jack. While IP telephones can be very simple devices, they are generally multibutton sets with an LED or LCD display panel. Some IP phones include an integrated Ethernet hub that allows the user to plug a PC directly into the telephone for a LAN connection. This approach minimizes wiring requirements. In some cases, IP telephones might include built-in Ethernet switches.

Because almost everyone uses a personal computer today, part of the trend toward convergence is device convergence. If a user has a PC, there may not be a need for a separate telephone device on the desk. Today, most PCs come equipped with speakers, and laptops include a microphone. A headset can be used with almost any current-generation PC to provide *softphone* capability in the computer. While the idea of a single desktop device sounds logical, there's a problem in real implementation. Desktop PCs use windowing software that allows the concurrent operation of any number of software programs. It doesn't matter what operating system the PC is running, the greater the number of tasks in operation, the great the likelihood that CPU resources will become overburdened and cause degradation in the telephone service. Even the simple act of writing out a large swap file to disk can seize control of system resources. The solution to this has been to create a telephone in firmware, or *firmphone*. This approach adds a coprocessor card to the computer that is dedicated to the task of telephone service, thus circumventing problems by providing a separate processing unit.

Naturally, these IP Centrex solutions all operate on the LAN, typically Ethernet, on the customer premise. For companies still using older shared-media technologies such as hubs, the addition of IP Centrex or any IP telephony solution may increase the offered traffic load on the network to the point that conversion to a switched environment becomes a necessity. Given that switched Ethernet solutions have been available since the mid-1990s, this is not a technology problem, but may present a financial hurdle that some companies cannot overcome.

IP Centrex Benefits

Having spent a number of years designing PBX solutions, I've never been a fan of Centrex services in the first place. As a result, it was difficult to address benefits objectively when I first began working with IP telephony solutions several years ago. Traditional Centrex provides some value in some very specific instances, but has often been a solution that was overpriced and ill-equipped to meet the needs of a business, particularly when offering it put almost no new requirement on the telecommunications provider. Centrex was often merely repackaging or rebundling of services intrinsic to the switching systems deployed by the telco. Not always, but often. IP Centrex does, however, offer a wide range of benefits to the customer.

Some of these benefits are part of the evolution to Centrex services for companies that hadn't previously implemented either Centrex or a PBX. Many are tied directly to the feature set described earlier. Some of these features are not available to smaller companies who cannot justify a PBX. Many of the benefits may be tied directly to outsourcing. If the cost of administering a system is prohibitive, outsourcing to a qualified provider has inherent value, if there is a business case for the services provided. Again, the need for a business case arises—in fact, some very large organizations are using Centrex services because the focus of the organization is not technology or telecommunications. Any organization would be wise to consider sticking to core competencies, and telecom services is what the ILECs provide as their core competency. The ongoing responsibility for day-to-day operations, administration, maintenance, and provisioning (OAM&P) is the provider's issue. Scalability as an organization grows can be a benefit, by eliminating the need for ongoing capital investment to continually enlarge a PBX.

Reliability is a factor that's rarely overlooked, but outsourcing to a qualified provider brings a benefit that many organizations cannot effectively address. The telephone companies routinely monitor the health of their networks 24 hours a day, 365 days a year. For an organization that has staff in multiple time zones, even an eight-hour day may be challenging to cover. The equipment may not truly be any more reliable in the electronics than equipment purchased and installed by a customer on the premise. But measuring the mean time to repair (MTTR) a problem requires includes the mean time to diagnose and isolate a problem. Telecom providers are, on the whole, far more capable in identifying and repairing problems than a company whose core business is in some other industry. That said, don't overlook that a good many outages in the telecommunications network are the result of construction work and backhoes completely outside the control of any provider.

Many benefits are tied directly to the integration of IP: convergence, convergence, convergence. Combining the existing voice and data requirements onto a single network infrastructure can have a huge impact on a company. Recent review of the billing for a relatively small business (about 120 employees in five locations) produced over 100 pages of monthly billing documentation when all the services were stacked together. Not just the technology convergence, but also the simplicity of dealing with a single network and single vendor can be a measurable benefit.

Multilocation Centrex is an approach that has been difficult at best in the traditional Centrex environment, particularly when two sites weren't in the same calling area. IP eliminates that problem and makes multilocation Centrex a realistic solution. Remote offices, small branch offices, even telecommuters can all be members of the same Centrex group. This approach allows for complete uniformity across the company with unified management of a single system. Services and dialing plans can be standardized and uniform at all locations.

Adds, moves, and changes to the system are the bane of every system administrator. With the implementation of IP telephony, the telephone number remains the same, while the IP address may change. No administrative work has to be done when a user moves to a different location.

Computer telephony integration (CTI) solutions become more practical than ever before. With the implementation of softphones or firmphones in PCs, the workstation can become a single unified interface to all corporate systems, voice and data. No special interface cards are necessary and the PC can eliminate the need for a telephone. This approach may be particularly beneficial in call center environments, which we explore in the last chapter of the book.

IP Centrex can provide all of these benefits without requiring a customer gateway on the premise. Figure 9.3 depicts this sort of implementation, which is more heavily IP based at the customer premise. The one new component shown here is a *terminal adapter* to allow connection of a standard analog fax machine to the IAD. This might be required in some configurations to support the analog telephone line technology over an Ethernet LAN.

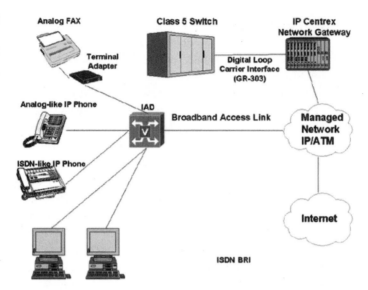

Figure 9.3
Integrating IP Centrex with the LAN.

Many people question whether this should really be called IP Centrex, bringing with it the negative past connotations associated with Centrex services. The idea of calling this solution a "Hosted IP PBX" or some variant has become popular recently. With the focus on Web-hosting and data center-hosted services using the ASP model, just treating it as an entirely new service could provide the sort of facelift the telephone companies need to make some solid inroads into the service market. On the other hand, if your friendly telco sales representative comes to sell you some new hosted IP PBX solution, you can ask them to differentiate it from any other IP Centrex service. The point is an old one...*caveat emptor*. Let the buyer beware—not of being sold snake oil, because that isn't an operating tactic any reputable telecommunications provider would use, but of being sold the wrong solution. An informed buyer is a happy buyer. It's up to you, the customer, to know what you need, and what you're getting. If you rely on your service provider, you will only get either as good as the sales team knows how to design, or whatever solution is paying high on the compensation plan that month. Know your business and your business requirements.

Does IP Telephony Include the Internet?

It's a good time to point out that every illustration in this chapter has shown a managed network using either IP or ATM connected to the Internet. In no case did we explore transmitting the IP telephony directly over the Internet. Voice over IP and voice over the Internet don't need to be tied together, and IP telephony shouldn't necessarily be construed as Internet telephony.

IP telephony might exist only at the premise running on an Ethernet LAN. It might run over a corporate Frame Relay network that's transporting IP. It might run over an ATM network, although those have not been widely deployed across a large customer base. The IP telephony network might well run over some IP, Frame Relay, and ATM combination using MPLS to provide a managed VPN solution. One of the greatest benefits of IP is its universal fit with many other technologies.

Using the Internet to transport telephony today may be asking for performance problems, because universal quality of service is not assured. Perhaps as the Internet continues to evolve and the PSTN and Internet merge closer and closer together, Internet telephony will become a commonplace reality. For the moment, it works...or not. And mission-critical business applications will always require appropriate resources.

Privacy: Fact or Fiction?

One area of concern in integrating voice service and the Internet is privacy. The perception, true or not, that most people have today is that telephone calls are private, whereas the Internet is fraught with privacy concerns and potential security breaches. The reality is that neither technology inherently offers more privacy than the other, and real privacy is a myth.

The reason most calls aren't monitored is that they have little information of value and tapping a telephone call isn't terribly easy. Even law enforcement agencies require a court order for wiretapping. The bottom line is whether you trust your service provider to deliver adequate protection. In the PSTN, there are a few very well-known providers. While there are certainly more since the breakup of the Bell System, the providers are well-known, reputable companies. The Internet introduces a problem because of the high volume of Internet service providers. A few thousand ISPs are active in the United States, far more than PSTN telephony providers, because the barrier to entry into the ISP market is relatively low in comparison to entry in the traditional telephony market.

Telephony providers have tools to monitor the PSTN; Internet providers have tools to monitor their networks. These tools are required to maintain and service either network. In the case of the Internet, some of these tools are the same as those used to maintain LANs, and they are readily available. Only two things are necessary to monitor network traffic: some form of packet analyzer or sniffer and access to the network.

Access is where the trust factor comes into play. A user can't plug into the Internet and monitor traffic across the continent. Typically, access is required on the segment being monitored. All reputable service providers monitor their networks for service conditions, but none allows outsiders, competitors, or anyone outside trusted staff to have the network access required to perform this kind of analysis. If you trust your provider enough to give it your business, you're dealing with a reputable service provider who is probably taking far more care to protect the network from intrusion than you might expect.

In the Internet in general, numerous cases of credit card information theft have occurred, but every documented case has been due to unauthorized or ill-gotten access to a database stored somewhere. No single documented case we know of involves a credit card number's being intercepted and stolen during transmission. If the Internet has proved safe enough for this use, and based on e-commerce revenues it has, it is safe enough for voice traffic when the time comes that network performance and quality issues allow voice transmission directly over the Internet.

Voice Over DSL/Frame/ATM— Variations on a Theme

Although the focus of this book is IP telephony, we would be remiss not to briefly discuss voice over packet technologies in a more generalized sense. Since many readers of this book are small business operators who are focused on DSL technologies, we briefly discuss voice over DSL, but the same principles apply to any packet-, frame-, or cell-based technology.

We've seen two things change the face of telecommunications in recent memory, deregulation and technology. (Perhaps given the most recent industry events, we should include accounting practices, but that's an area for some other author to explore.) Some of the most dramatic changes in telecommunications have taken place in the local loop. Deregulation and new local loop technologies such as DSL have created opportunities for the incumbent providers, and created whole new industry segments that didn't exist in the past. Voice over DSL (VoDSL) represents one of the new opportunities these changes have created.

While estimates place voice and data network load at comparable levels, nearly 90 percent of telecommunications revenues are derived from voice services. Census data indicates roughly 8 million small businesses in the United States.

In 1996, the U.S. Congress passed the Telecommunications Reform Act. This opened the network to local competition by requiring ILECs to make some network elements in the PSTN available to competing service providers. The most obvious of these elements was the twisted pair local loop. TA-96 laid the groundwork for competitive service providers to install transmission equipment in the local telco CO and connect it to the local loop. This act has been a driving influence behind many of the competitive DSL service efforts since 1996.

One of the benefits of DSL in today's environment is that it supports the use of the telephone at the same time as the data connection. It isn't IP telephony, but it's a step towards convergence, and it does place both traffic types on the local loop at the same time. Another variation is sending voice traffic within the DSL frames themselves. This telephone service is called *lifeline service*, due to the quality available, but also because the telephone is a "line-powered device," which operates in the event of a power failure.

Voice over DSL (VoDSL)

Voice over DSL (VoDSL) is a technology that provides for the transmission of voice traffic using the actual DSL frames to carry the conversation. The DSL family comes in a staggering array of different technologies. We only refer to *asymmetric DSL*, which might be called ADSL, g.LITE, Consumer DSL, or by some other marketing name in the United States. This is the most widely delivered consumer DSL technology. It can support multiple calls over the telephone line and takes advantage of the DSL framing technologies and line-coding technique known as *discrete multitone technology* (DMT).

DMT allows for multiple voice-grade channels over the twisted pair within a frequency spectrum ranging from 0 Hz to about 1.1 MHz using standard V.34 modem technology. This technique can provide 33.3 kbps throughput from a telephone—the standard 4-kHz voice channel we studied earlier. DMT works in ADSL deployment by dividing the frequency range between 0 Hz and about 1.1 MHz into 256 subchannels or "bins," each of which is the equivalent of a voice-grade channel, and is capable of 33.6 kbps in throughput. V.34 modems allows channels to be bonded together, but this technique requires provider support and is rarely implemented. ADSL uses this same technique to bond the channels together in a form of inverse multiplexing on a single chipset, thus providing the high-speed bandwidth so popular in DSL solutions today. Each is a 4-kHz voice channel, but operating at different frequencies.

Figure 9.4 demonstrates in general terms how DMT line coding makes use of the frequencies used by DSL. The first channel or bin is the actual voice circuit used for POTS. Above that is a *guard band* that remains unused to provide for some separation between voice and data services. The subchannels used for ADSL start around 25 kHz and continue from there up to the highest frequency used. This approach is how DSL provides dynamic bandwidth capability and is able to adjust to varying line conditions. It also explains why different users may measure different throughputs based on current conditions on the local loop.

Figure 9.4
Discrete multitone line coding.

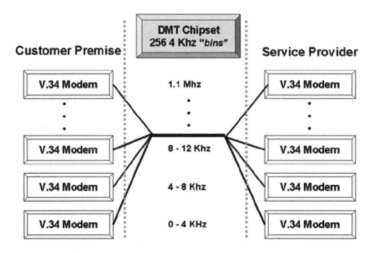

If the DSL bandwidth isn't being used for data transmission, the extra capacity can be used to carry voice traffic inside the DSL frames. This approach brings the voice and data services even closer to being a single converged service. Naturally, the service provider's profits are increased by leveraging DSL access lines to provide both voice and data services. This approach is sometimes referred to as *derived voice* because a virtual voice channel is being derived from the DSL traffic stream. Because DSL uses the bandwidth dynamically, voice calls use the bandwidth only when necessary.

From a business standpoint, this approach can consolidate service to a single billing arrangement with one provider. This single point of contact can be a tangible value. This solution uses the existing loop technology, but requires VoDSL equipment installed on the customer premise. The VoDSL approach is shown in Figure 9.5. One of the advantages shown is the ability to use standard analog telephones and fax machines, which are supported by VoDSL equipment. This equipment is also available in configurations to accept other types of connections.

Figure 9.5
Derived voice over
DSL (VoDSL).

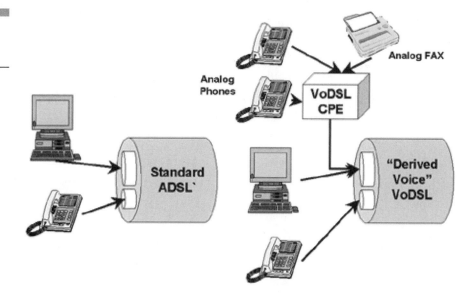

Several combinations and configurations can permit different approaches to providing this service, but they still all rely on the availability of DSL, and most frequently VoDSL support must come from the DSL provider, which is primarily the local exchange company. LECs provide this service offering, and DSL in general, focused on extending the life of their equipment and providing necessary service to protect their embedded customer base for voice services. For businesses looking to broaden their focus to other service providers beyond the traditional telco, these solutions may not meet future requirements.

Services Perspective

Overall, any managed service is worth evaluating for your business' individual needs. If it adds values in some quantifiable way, or provides a service you cannot provide yourself that is important to your business, it's worth considering. In general, if providing the service means adding a skill set to your company that isn't currently present, outsourcing may be the right choice.

IP telephony service offerings are primarily those described here, coupled with managed network offerings, service level agreements tied to committed quality of service and delivery, and in some cases, IP-based VPN solutions today. That will not remain the case for long. With the current struggles of the telecom sector, increased new offerings and competition will appear in this area. IP telephony has long been identified as a key disruptive technology in the telecommunications industry. Managed service offerings are popular and growing in their popularity, which might make a managed IP Centrex offering the quickest path to a fully converged voice and data network for many companies.

The Future
of IP Telephony

IP telephony holds great promise for the future as a cost reduction measure, but most truly as a means of creating converged network infrastructures running multiple application services. We've seen a tremendous migration of the intelligence of the network in the past few years. In the mature PSTN, all the intelligence lives within the network. The end devices are telephones that have no CPU or intelligence of any kind. This environment really turned the customer into a commodity, and telephone companies focused on how many their local loops served. A direct billing connection with the customer was driven by the customer's dependence on the network. In the new Internet world, intelligence is migrating to the end stations in computers, but also to the edge of the network. Services are moving closer and closer to the user, but with dynamic IP addressing, the location of the user is irrelevant. The user doesn't have to be at a specific device or location in the new network model.

One necessity for the continued growth and success of IP telephony is a steady advance in consumer broadband solutions. The issues of the last-mile delivery of services have been the topic of raging debate for far too long.

DSL and cable modem are essentially legacy technologies, reliant on a predominantly copper local loop. Hybrid fiber-coax has aided in the advances of some cable modem solutions, but these options are both the past, not the future.

- DSL will never succeed until TA-96 regulations are modified. The ILECs have a strong incentive not to enter into widespread deployment of consumer DSL under current regulations. The Tauzin-Dingell Bill (HR1542) could impact that, but is poorly written, and unlikely to pass the Senate in its current iteration. It is not universally supported by the telecom industry.

- Relaxation of regulatory requirements alone will not assure DSL deployment. Since divestiture in 1983, the ILECs are companies with a mandate to be profitable. They have to make a legitimate business case for return on investment to deploy a service en masse. DSL often does not meet business criteria for good business. In many cases, it was deployed as a knee-jerk defense to cable company encroachment into the ILEC embedded customer base.

- DSL is too little, too late at too high a price for everyone except consumers in very densely populated areas. At best, it may have a usable life span of 5 to 7 years. If DSL ever succeeds, it will only be to extend the life of the local loop and further protect the investments made long ago by the ILECs.

Both voice and data networks are using fiber optics in the core and middle mile, with a wireless connection to the customer. This next generation of the Internet is shown in Figure 10.1. We show a "service POP" for the local provider. It's clear that this doesn't need to be an ILEC. In many cases, it will be an ISP, a CLEC, or some new blended provider offering voice, data, television, and any number of other possible services to subscribers. The network will use optical technology like dense wave division multiplexing (DWDM) to reach subscribes over *fiber to the neighborhood* (FTTN). The local loop will sometimes be a fiber-connected loop, particularly in the case of office buildings and business parks, where large businesses need to move massive amounts of information. For most consumers, the local loop will become a wireless connection.

Figure 10.1
The future of networking.

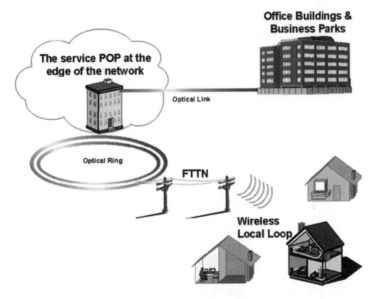

802.11 wireless is extremely popular at present, yet no major players are deploying it on any great scale. There are two clear reasons for this. First, the spectrum remains unlicensed at present. That poses issues for widespread commercial viability. More important, the incumbent rarely has incentive to lead the way to a disruptive technology. Why would the ILEC move to a technology that can so clearly eliminate any revenue stream from that twisted pair local loop?

Friendly legislation and support for deployment scenarios would ease the migration to newer technologies and provide incentive for the incumbents to give serious credence to the technologies. Legislation should foster competition, not discourage it, but good legislation must foster fair and equal competition, something we've not always seen due to political and financial influence.

Advances in optical technology and a glut of fiber have conceptually turned bandwidth into an infinite commodity, but the reality is very different. Unfortunately, the advances being realized in the core of the network have been terribly slow to pass on to consumers. DSL-like speeds at dial-up prices are the only general solution consumers will embrace.

Beefing Up IP Telephony

IP telephony has emerged from a hobby-quality niche to a competitive, viable solution that is being accepted more widely every day. The progress made over the past two years brought the technology into the "carrier class" service category, and many carriers are working at integrating IP networking into the telephony core of their networks. Areas still exist where development work is needed for the solution to grow to a healthy replacement for traditional telephony, but equipment manufacturers, standards organizations, and other players in the industry are addressing every issue identified.

Personal computer processor power still poses potential problems in real-world implementations. The typical PC sold today comes equipped with a 1.2-GHz Pentium class central processor, 128 to 256 megabytes of RAM, and a spacious hard drive. But the most common operating system in use is the Windows family of working environments, and all remain terribly processor intensive. Running H.323 has already been described as overhead intensive. SIP is far more "resource friendly," but the fact remains that using a PC to make telephone calls can still result in unacceptable performance degradation without the support of a coprocessor card. Certainly, one solution will see a more integrated telephony solution provided by Microsoft, but that isn't the best way to resolve the performance issues. A more robust approach would be to enhance PC hardware to include hardware coprocessing capabilities for encoding and decoding voice traffic. This would allow Linux and other future operating systems to include universal support for telephony. A standards-based *telephony application programmer's interface* (TAPI) can provide tools that allow tighter integration of all software. The abili-

placing the telegraph system with a completely new, real-time, interactive communications tool. But no technology can remain disruptive. Once it becomes mainstream, as the PSTN did long ago, it becomes a sustaining technology. And once the sustaining technology matures, incremental advancements slow, and the door to disruption from outside opens.

Let's perform a quick side-by-side comparison of traditional telephony and IP telephony and see just how disruptive this solution might be; see Figure 10.2.

Figure 10.2
Mainstream telephony versus disruptive IP telephony.

Traditional Telephony PSTN	IP Telephony VoIP
Deeply entrenched in mainstream PSTN	Overlooked as unnecessary by many PSTN providers
Well established value model	Requires a new value model
High performance	Lower performance
Low cost due to years of experience	High cost to deploy. New systems, training, methods
Large established market	Unknown and untested market
Large companies cannot ignore the technology	Large companies can easily ignore the technology
Mature, well-developed products	New, evolving products

IP telephony pretty clearly sits in a position to seriously disrupt the telecommunications industry. Called minutes billed to IP telephony are rising at dramatic rates, with expectations of an ongoing multibillion dollar revenue stream. Over the past year or two, ongoing advancements have moved IP up the market chain, making it more and more viable and worth serious consideration.

Instant messaging (IM) applications play a role in this disruption activity as well. This tiny application has become one of the most widely used utilities on the Internet. As companies discover employees using IM in productive business ways, they will embrace it more and more. There's a fairly tight synergy between IM and telephony, with the ability to shift from short messages by simply clicking a button to speak to someone. These applications will converge to some degree over the next year or so as both continue to mature.

IP telephony services and applications are poised to move quickly from a minor, yet significant portion of the telephony market to a major and growing share. As the existing providers struggle to find a new identity in the market, and fight to regain consumer and investor confidence, the providers that only offer IP-related service have an opportunity to seize major market share. The advancements in features and technology are continuing at a rapid pace. The products are maturing, and while they haven't fully matured, an analysis of cost versus benefits will often tip the scales solidly toward IP telephony as the best solution for many businesses. That trend will continue, and IP telephony will very likely become at least equal to traditional telephony in the most technology-oriented markets (North America, Europe, and much of Asia).

This is a pretty clear wake-up call to the telecommunications industry: there's a barbarian at the gate, and it's called IP telephony. As users find new ways to implement and use the technology, this particular barbarian could overrun the city, leaving few profitable survivors in the traditional telephony sector.

Internet Call Centers

One implementation of IP telephony that will grow in the next year is the *call center*. We've seen many changes in how people work; more and more people focus on family and friends and want to change how they participate in their work environment. Internet call centers provide an approach that offers a work-at-home job, but with far more connectivity than we've previously seen.

Call centers have historically driven jobs to areas that offered a lower tax base to businesses. Many U.S. companies now use offshore call centers in other countries, where the labor rate is lower; however, the distributed call center approach alters the cost structure and provides an effective method to hire domestic staff around the country. With the growing e-commerce environment of the Web, call centers have proved to be a growing segment of the market, yet traditional telephone technologies still present a barrier.

The leading approach for home agents has historically been ISDN. The integrated services digital network still doesn't provide integrated services. The bottom-line costs wind up being too high for most companies to invest in this architecture. A shift to IP telephony reduces the cost and provides better integration of services than ISDN has ever offered. Many companies are now looking at IP telephony, coupled with DSL and cable modem solutions, to fill future call center requirements.

Perhaps the greatest vision of promise in call center technologies was driven by the burst of the dot-com bubble. A large number of companies tried to enter the Internet market and sell all manner of goods on the Internet. These businesses had no "brick and mortar," which was one of the attractions, but many also had no history in providing customer service. Some succeeded, but even Amazon, one of the most successful examples, has yet to become a profitable company. Most attempts failed, and many failures related directly to the inability to provide proper customer service. The focus of any company doing business on the Internet has become customer service. The demand for customer service is best filled by the ability to speak with a live person. The call center, in any form, provides that ability.

Consider Figure 10.3; the telephone network and Internet are both represented, but linked together much as they are in the real world. Customers might be at home or anywhere. They might contact the provider via telephone or via a Web site. According to the Gartner Group, more than 70 percent of transactions take place over the telephone. Those Web sites that have been able to implement live voice support for potential customers report as much as 50 percent increases in sales. Online worldwide revenues from retail sales in the United States are anticipated to hit $35.3 billion this year.

Figure 10.3
Merging the PSTN and the Internet.

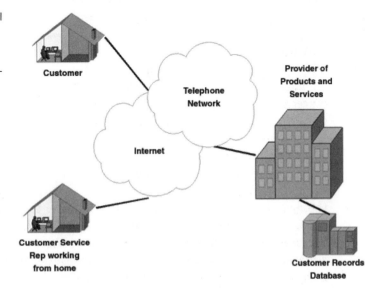

Customer

Telephone Network

Provider of Products and Services

Internet

Customer Service Rep working from home

Customer Records Database

The provider of products or services receives a query for customer support and through distributed call center technology, is able to redirect that call to a customer service representative working from home.

The staffing of call centers has always been a difficult issue. The ability to hire remote staff, perhaps even part-time remote staff, allows the provider, regardless of location, to find good, qualified employees. Time zones become a non-issue. Even customer service reps with special linguistic skills become obtainable resources. The benefits to the telecommuter work force have been studied time and again in the past 10 years or more, and everything suggests continued migration to work-at-home efforts.

This solution also provides an opportunity for a pool of workforce candidates that may have been inaccessible in the traditional call center. Stay-at-home mothers, retirees, even people without transportation now become potential job candidates, participating in the workforce in ways they may not have been able to previously.

Distributed call centers do not require IP telephony, but it does provide the greatest level of integration at the lowest cost. The distributed call center has often been implemented using PBX solutions and off-premise stations or ISDN lines. ISDN can be expensive, often more so than DSL or cable modem, and it is not at all ubiquitous. The newer generation of IP and Internet technologies make the distributed call center more cost effective to implement than it has ever been in the past. The use of IP technology as a "PBX extender" creates a virtual call center environment that can physically be anywhere. Employees needn't be passed over because of their geographic location. To customers, the company presents a single unified point of presence.

A broadband or high-speed connection capable of supporting voice and data simultaneously is optimal, with DSL the technology of choice for most companies implementing these call centers today. The "home office" or heart of the distributed call center is typically traditional call center technology but, as we've already seen, it is easily replaced by an IP solution.

Distributed call centers use a job performance technology that requires managers take a more "hands-off" approach to supervising workers than traditional workflow methods. Supervisors learn to rely on the systems, both telephone and computer networks, to measure and monitor productivity and worker activity. The idea of a worker in the corporate office where work can be directly observed is transformed into a measurement of productivity and results rather than activity.

For some companies, the challenge of remote teleworkers becomes one of combating isolation. Because workers don't have the social interaction with colleagues (i.e., water cooler chat), supervisors must adapt a management style that encourages interaction at all levels to minimize the chance that remote workers are left "out of the loop" and disenfranchised from the business of the company. Conference calls, computer video conferencing, and regular visits to the office or meetings with colleagues have proved effective in overcoming this issue.

The distributed call center provides a very attractive alternative for companies in large metropolitan areas that often have alternative commuter requirements because of air-quality management regulations. This solution can aid in bringing a business into compliance with these regulations.

Although there has been a trend to move call centers offshore, today many companies have become very security conscious and are more reluctant to engage in offshore arrangements. The distributed IP telephony call center allows for substantial savings above traditional costs without sending jobs outside the United States, although that option can also work using IP telephony.

The benefits of distributed call centers can be measured in a variety of ways. Some benefits cited by companies implementing this technique for managing a call center business include:

- **Reduced cost of office space**—Because teleworkers do not require a cubicle or workspace in the corporate office, that office can shrink, thus reducing real estate and associated costs. In reviewing real-life implementations, the companies interviewed estimate that building costs alone were recouped within 3 years by shifting to a distributed model.
- **Location**—From the mid-1970s through the 1980s, a migration of call centers from major metropolitan areas to more rural settings provided a readily available workforce and a reduced-tax incentive to business. In particular, Omaha, Nebraska provided incentives to companies considering relocation and successfully attracted many call center businesses to the area. Using the distributed model, corporate location is irrelevant, and so is the location of the teleworkers.
- **Tax benefits**—In major metropolitan areas particularly, mandates are in place that require businesses to encourage alternate commuting methods. Car pooling, bicycling to work, and telecommuting are all proven alternative commutes in the proper setting.

Any organization that interacts with a distributed base of customers over the telephone may have need for a call center:

- Healthcare providers, particularly large organizations and HMO/PPO groups, needing to handle patient calls
- Insurance carriers dealing with customer calls regarding policies
- Catalog or Internet-based merchants with a high volume of telephone transactions
- Airline, hotel, event registration, and ticketing agencies
- Financial investment firms, particularly those dealing in high volumes of telephone calls
- Social services organizations of many types providing call-in services for their clients

In general terms, the call center is probably not an appropriate technology for the small business. This solution is the best fit for high-volume, transaction-based services that require interaction with a customer service representative. On the other hand, a small business with an IP Centrex solution can implement call center services easily to provide new levels of customer service. In short, IP telephony brings the call center into the reach of a small company without the burden of a full-blown, expensive implementation. IP telephony not only disrupts the telecommunications industry, but also can level the playing field for smaller companies, providing tools that, in the past, have only been available to businesses with a large budget and staff.

Fax Over IP

As estimated 80 million fax machines are in use in the world today. In many small businesses, the fax transmission of orders is standard routine. Yes, when we think about fax traffic, it really isn't real-time traffic requiring a circuit-switched connection. Fax traffic isn't delay sensitive, at least in the sense that a few seconds delay won't damage the integrity of the transmission. The question is whether or not it really makes sense to tie up circuit-switched connections in the PSTN to send fax documents.

Matters are complicated further by the document scanning and transmission process. At the scanner, the fax machine reads an analog document from a piece of paper and digitizes (or rasterizes) it into digital form. Now that it's been converted to a digital bit stream, it's ready to send, but the medium we use to transmit fax is a plain old analog phone line. To send digital information over the analog line, we have to modulate it with a carrier using a modem. When it hits the local CO switch, it then has to be sampled and converted for transmission over digital carrier trunk facilities through the network. At the receiving end, a modulated carrier is sent out over the analog local loop, which the modem in the fax machine must convert back into digital data. This digital data is then used to recreate a facsimile of the original by printing out on an analog piece of paper. As you can see in Figure 10.4, this entails six analog-to-digital-to-analog conversions to transmit one page of information. Why not digitize the document once and complete the whole process digitally?

Figure 10.4
Analog-to-digital
conversion for fax.

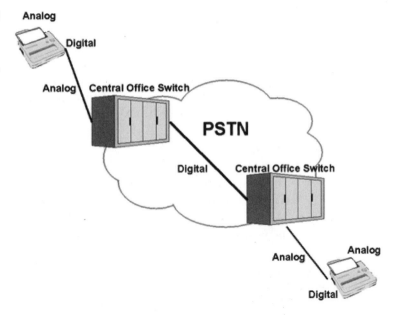

In many cases, we avoid the fax machine entirely and send documents as digital attachments in email. When we do this, we eliminate the conversion process and the need for circuit-switched resources. But we don't always have an electronic version of the document we need to send. Sometimes it's a copy of a magazine article, a sales brochure, or a preprinted contract. Since we're spending several billion dollars a year sending this sort of information, there's clearly an opportunity to take advantage of the technology.

New solutions reduce this problem: fax servers and fax gateways on LANs are more and more commonplace; and fax-service Web sites are very popular and becoming more so. (Readers are encouraged to look at www.efax.com and www.jfax.com for examples that can provide packetized fax service, which not only provides saving on paper manufacture, but provides a means to move fax off the PSTN and transmit via the Internet.)

There is still room for enhancement in IP fax technologies. This is an area where billions of dollars are spent every year, as much as 20 percent of toll and long distances revenues. The opportunity for cost savings is huge for companies that transmit or receive high volumes of faxed material.

Voice Mail over IP

Voice mail has become such a commonplace tool that many people take it for granted. The days of tape drives to store messages, even on answering machines, are long gone. Stored voice mail messages are now digitized voice stored as data files on some form of memory device. In corporate voice mail systems and telco service provider networks, messages are stored on a hard disk.

One question, as the Internet and IP telephony progress, is just how many voice mail systems do we need? How many do you have? Many people have voice mail at home and at work, but then there's the home office. And what about voice mail as a standard cellular service? The technology issue is that, given multiple voice mailboxes where this information is all stored in digital form, how many telephone numbers does the user need to call to retrieve messages? Two? Three? Four? And why? Why not just build a user-programmable interface that delivers the digitized file to a central point for retrieval? Why not let the user program in an email address and deliver messages to email as an audio attachment?

Like fax service, voice mail delivery isn't delay sensitive. The message needs to be delivered intact, but even a few seconds delay in delivery won't harm message integrity, as long as the entire message is delivered. Voice mail isn't an interactive service and need not be treated the same as real-time traffic. Like other forms of data traffic, it is sensitive to errors, but not to delay.

To some people, this still sounds a bit farfetched, but readers might wish to surf the Web to www.onebox.com or www.address.com to see live examples of just this type of service.

Several streaming audio technologies work very nicely at delivering this sort of information, both proprietary and openly published. Anyone who has ever listed to Internet radio has heard the technology. It may not be interactive voice quality, but with some buffering, even dial-up modem connections to the Internet allow for the delivery of voice mail messages using these protocols and techniques.

Unified messaging is discussed, but not truly available...yet. It will be. This service offers promise in the next cycle of advancements to IP telephony–related technologies. This advanced next generation network of IP telephony is shown in Figure 10.5. SIP and H.248/Megaco are the control protocols within the IP network and operate between the call agents, the servers, the trunking gateways, and the integrated access devices (IADs). Media streams are handled by a combination of RTP, MPLS, and other protocols (like MPLS in conjunction with ATM) between the routers, gateways, and integrated access devices. SS7 is used between the Gateways in the IP network and the PSTN. Connecting lines have not been included for all of these relationships because the illustration would become too busy, but as you envision each, you can see that the next generation IP telephone network is a very sophisticated and complex environment.

Figure 10.5
Next generation
IP telephony.

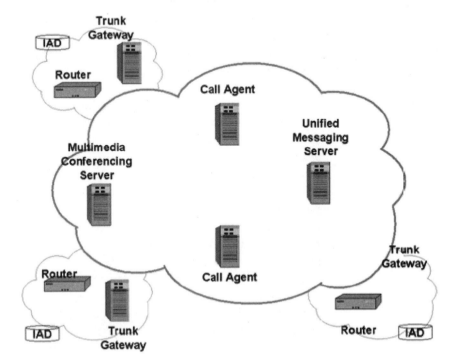

IP Telephony and Denial of Service (DoS) Attacks

Some concern exists in the Internet community about a recent series of problems referred to as denial of service (DoS) attack. This discussion is not designed to convince anyone that DoS attacks are not a real and growing problem.

In the IP network, every device has an individual address that allows it to become a fully participating member of the network. This host address is unique for every connected device. In the Internet in particular, addresses are widely known, even advertised in some fashion. The result is that if someone with malicious intent knows the IP address of a potential victim, it is possible to flood the victim's address with more traffic than can be processed, thereby denying service to legitimate users. DoS attacks are often much more sophisticated, but for purposes of discussion, this simple description is adequate.

The danger to the Internet user is that often there is nearly no defense against denial of service attacks. If a company needs a public Web site, that site must have an address users can reach. The network has no way of identifying undesirable traffic, thus making it very difficult to eliminate threats before problems occur. It's difficult, but not impossible. Many intrusion detection systems, firewalls, and routers can detect some predefined conditions that may indicate an attack is in progress and take defensive action. Nonetheless, the bottom line is that any host connected to the Internet is a potential victim of this type of attack.

Denial of service attacks are further complicated by the attackers' tendencies to use falsified, or *spoofed*, IP addresses to hide the true source of the attack. Attackers have gone so far as to initiate *distributed denial of service* (DDoS) by loading a program onto remote computers that allow the attacker to seize control. These programs, referred to as *zombies,* can even allow an attacker to use an unwitting victim's computer to attack a third party.

Although this sounds like a potentially fatal flaw in using the IP environment or the Internet for telephony, that simply isn't the case. It has been argued that users on the PSTN can't be blocked from service in this manner, but caller ID information can be used to block the offending caller and prevent blockage of the user's telephone. And while this is true enough, it's a reasonably safe assumption that given a large enough bank of pay phones, rolls of quarters, and accomplices, denial of traditional telephony services is actually quite achievable.

Detractors speculate that this vulnerability presents an insurmountable barrier to IP telephony and makes the solution no longer useful. Not really. It presents a challenge and a responsibility to network designers and administrators. Designers are responsible for the intelligent, fault-tolerant design of networks. Networks cannot be allowed to just multiply and grow without planning, as they did in the past. Network administrators must also embrace the need for closer monitoring and auditing of network traffic. No longer can the network be turned on and ignored, with the assumption it will run error free. The network is no longer merely a simple Ethernet LAN, but a complex organism providing multiple services and requiring much of the same attention to security, traffic engineering, and monitoring as the PSTN has long required.

The Future for Equipment Vendors

There's really only one guiding principle for equipment manufacturers to focus on for the future of IP telephony: innovation. Don't duplicate; innovate. Incremental improvements are expected and necessary, but don't expect them to sustain an equipment product line for years. Issuing the next "dot release" of software that adds two new features isn't innovation; it's stagnation. Push for new ideas and encourage an entrepreneurial spirit within your company.

Even the largest, most widely accepted products are subject to disruption. Several vendors have already seen the impact of leaps in improvement by competitors. Exponential advances quickly become incremental sustaining advances in technology. Research and development work must be encouraged and cultivated, even knowing that the old funding models for such activity no longer work. Venture capital for speculative efforts won't come easily. Cultivate small pockets of innovations within your company. And don't punish mistakes. They will happen as part of the development process. To paraphrase Confucius, he who doesn't make mistakes, doesn't make anything at all.

Let's consider a simple theory in technology equipment: acquisition speeds the path to stagnation and stifles innovation. The trend of the past ten years is to let small companies innovate, then, as technologies prove themselves, to purchase the innovators. It sounds reasonable, but there's a two-fold penalty to pay for this approach. First, the new technologies come from different processes and may not integrate well into your existing product lines. We've seen this scenario played out time and time again as a vendor's product line identity shifts, trying to accommodate new solutions that were developed with a different approach. A far more dangerous penalty comes from the message this sends to your own internal team: we don't innovate here; we look to the outside. This approach stifles the creative spirit of your staff and, in the long run, motivates creative designers and entrepreneurs to move on to other opportunities. Given the speed of technological advancements, driving innovation out is the surest way to propagate mediocrity. It might not hit today, but it's a sign that a company is falling into the sustaining technology mindset and becoming a candidate to be overrun by another disruptive technology.

Equipment manufacturers no longer make just one or two components. Now, single vendors provide the router, gateway, softswitch, integrated access device, the LAN hubs and switches, the firewall, even the telephones. Manufacturers must attend to not only the design and management of their systems, but to the simplest things. Standard configurations in this equipment should reflect a reasonable and realistic approach to design. Make the defaults more like what customers really do. Make it easier to fine tune configurations to fit the implementation. Provide simple voice services in even the low-end products, and simplify their configuration everywhere possible. Simplify and integrate more tightly, while adhering to open standards of interoperability.

Don't give your customers a gooey GUI. So many vendors think the more complex a GUI is, the more complete it looks. Some products in the market appear to have more development work applied to the user interface software than to the actual product, so here's a word of caution. Treat your GUI interface as if it were the Web. Two or three clicks should take a user to where they need to be quickly and easily. Don't make customers hunt through layers of menus and nonintuitive options trying to find out how to add a user or other simple task. Simplify!

Customers don't want to buy more of the same thing. For equipment vendors, telecommunications carriers and ISPs represent the largest customer segment, so look not to their success, but to their failure in technology. Give them solutions that do the things they can't do today. Resources galore identify where end user needs are not being met. Long-term success is not found by creating new business models, but by creating the products that serve customer needs.

The Future for Telecommunications Service Providers

Like equipment manufacturers, service providers must innovate. The telecommunications service industry has become a gargantuan, lethargic beast. It's true that we've seen large major carriers undertake widespread deployment of cellular services to provide telephone connectivity anywhere, and we would argue that cellular services are sustaining advances in the traditional telephone service model. They're an improvement, an increase in availability. They may even be something of a large step, but it's still an incremental large step. Cellular service in the United States has become a sustaining technology, and new advances are incremental chest-thumping attempts to gain market share. While integrating every square foot of the United States into a coverage area improves the service and increases the customer base, it isn't innovation.

Managed service offerings hold great potential for the future. Today we see managed VPNs, managed firewalls, and managed networks. Web-hosting services are offered, but in most cases they are still immature. Providing a building with power, fiber access, and a secure environment isn't really hosting—it's real estate management. Sell services for what they are, and explore the new requirements. Offer a managed data center integrated with remote hosted IP PBX services that scales from a company of 10 employees to a company of 100,000 employees. Manage it wisely and well. Don't overcommit. Don't sell smoke and mirrors. Build a valuable managed service by asking customers what they need, then provide it. Building solutions based on what looks profitable and manageable creates nice, clean service offerings that fail to meet customer needs.

Speed is of the essence. A recent implementation of a major provider's offering of a managed service for fewer than 30 locations took nearly one year from initial proposal to cutover. This wasn't a new offering, but one that is widely deployed. This plodding, inefficient process hurt both the provider and the customer. The provider didn't see any revenue from the new service for nearly a year, but invested tremendous resources in the interim. The customer didn't gain any advantage for the whole timeline. Speed and efficiency in providing managed services are critical to their success. Providers must demonstrate their proficiency at managed services or lose customers to competitors.

Telecommunications carriers need to look to small competitive ISPs and CLECs. See how they're succeeding and take advantage of their efforts. Take the things they do, do them better and improve upon them. Your biggest disadvantage may be your size. A corporate telecom elephant may find dancing the jitterbug very awkward. Foster a corporate culture that is lean, mean, and responsive. Your smaller competitors will always be faster and probably always more innovative. Don't rest on the laurels of your embedded customer base and think that you can catch up later. That's living in denial. Later is too late.

The incumbent telecommunications carriers in particular should recognize DSL offerings for what they truly are: a last-ditch effort to maintain the customer relationship over the twisted pair. It's time to accept that the twisted pair is a legacy of the past. Hang on to the copper, and you're tied to the past. It's time to invest research and development efforts in wireless technologies. Some wireless solutions available today have been tried and proven questionable. Look at them again and re-evaluate service deployment models. Investigate 802.11 WiFi technologies, not for carrier class deployment but for what it is—a simple LAN technology. But the groundswell of popularity is rapidly proving that it meets customer demands. Look to the future rather than the past.

ISPs and CLECs need to look to the telecom industry. Service level agreements and guarantees of delivery are areas that Internet technologies have always been weak in. The major telco carriers have an advantage in their large infrastructure, because they can often provide services that a smaller company struggles to compete against. Your strength is your size. Smaller companies tend to have an ingrained competitive and entrepreneurial spirit that encourages new ideas and new thinking. Seize the hot new markets quickly. Work closely with your customer base to give them what they need before they need it. Understand and anticipate your customer's requirements before they do, and you will win customers faster than you can imagine.

The convergence of voice and data networks is a reality. It is what customers want. Give customers what they want to succeed. It's a very simple business tactic. The next generation of wireless, integrated tightly with IP telephony, will not only disrupt the traditional telephony market, but the cellular market as well. (The wireless PDA has struggled to find solid acceptance.) When we think of the Internet, we tend to think of the World Wide Web, but there's no reason that information can't be accessed via other methods. Couple IP telephony with wireless networks. Tie in instant messaging. Integrate data query using location-based services. Yes, it sounds a lot like 3G wireless, but consider how disruptive it is to the traditional telecommunications industry.

The Future for Business Users

This book is about a technology, but the technology is merely a tool that brings us a step closer to full convergence of voice and data networks and services. Let's consider just why this type of convergence is important to business.

The two most valuable assets a company has are its people and its intellectual capital. Intellectual capital might be housed inside network servers and data center archives, but it's often most commonly found inside the staff of employees. Data, information, and knowledge are more valuable that inventory and cash reserves. Yet even this mindset fails to account for human capital. The information model in Figure 10.6 takes this concept one step further.

The raw data is easily turned into information that tells us Ken is 6 feet tall, lives in Vermont, and has a dog named Zoë. Even automated systems can incorporate logic rules that can extrapolate knowledge from this raw data. We know Ken is easier to reach by phone than email because we have a telephone number, but no email address. Wisdom is only gained through the human factor. Only personal conversations on a human level discover that Ken takes Zoë for a walk in the woods or to play in the lake in the late afternoons. This *wisdom* cannot be extracted by machine. It requires the one computing device that can think extemporaneously, building relationships between data elements dynamically...the human brain.

Figure 10.6
How raw data
becomes knowledge.

DATA	Ken, 6, Brown, Vermont, Dog/Zoe 802-555-1234
INFORMATION	Ken is 6 feet tall, has brown eyes, lives in Vermont and his a telephone number is 802-555-1234
KNOWLEDGE	Ken is easier to reach by phone than by email
WISDOM	It is best to not contact Ken in the late afternoon

Raw data isn't much use if you don't know what to do with it, but a good system of sharing knowledge and wisdom can be the secret to better efficiency and higher profits company-wide.

Raw data is easily collected. In the past, it often went into archival storage, buried in a tomb from which it could never be retrieved. The past few years have focused attention on *data warehousing* concepts, which rapidly evolved into *knowledge management solutions*. Wisdom remains in the human domain, outside the reach of computers.

For a business to take advantage of the wisdom in human capital among its staff, that business must give the staff tools and knowledge upon which decisions can be based. The closer at hand the tools are, the more efficient and productive a company will be. To bring the tools closer to hand, create a converged environment where access to any piece of information and any communication need is available instantly.

Business customers have a great challenge, because new technologies cannot be overlooked. They must be evaluated, and a business case analysis performed, but the analysis must be done swiftly. Implementations must be quick and efficient. Don't place unwarranted faith in an equipment vendor or provider. You are managing your business; they are managing their business. Insist on performance commitments and service level agreements. Don't assume you're getting what you're paying for. Monitor your systems and know what you're getting.

Managed services may offer the best hope for many businesses. Evaluate them closely. IP Centrex clearly needs to be called something else. Coupled with data center hosting services, it provides the truest service convergence a provider can offer. Today, the offerings are limited, but they're improving. Work closely with service providers. Above all, be vocal. When offerings fall short, make the shortcomings known. The world of telephony is far more than a dial tone now. Don't settle for a dial tone.

A converged single network creates perhaps the greatest differentiator between competitors in business. Build your business with an eye to the future. Incorporate technologies directly into your business when they fit with the core competencies. Outsource to qualified providers when appropriate, but manage the providers lest they manage you. Give employees the tools to do the best possible job in order to receive the best possible results.

The Evolution Continues

Humanity has evolved through several different eras. Nomadic tribes of hunter-gatherers followed the migration of animal herds or the availability and growing seasons of crops. Over time, this gave way to the

Agricultural Age that gave rise to cities and towns (or castles and villages), which were later connected by roads. The Industrial Age brought the automobile, tying cities and people closer together. It also focused on productivity in the manufacture of goods. The Industrial Age was an age of commoditization.

While all of these will forever remain important, the age of information is well upon us. Data, information, and knowledge comprise the intellectual capital that makes up the value of much business today. Manufacturing processes have become so efficient that goods quickly become commodities, but the services associated with those goods also bring value.

The telecommunications industry, as we've seen it grow and mature, is now gone. It will never be restored. It's perhaps important to put Darwin's theory of evolution in perspective. The theory never suggested survival of the "best." The theory of evolution really describes how the most adaptable species will survive and rule the day. To merely survive, we must be nimble and adaptable in embracing the Information Age and all its technological challenges. To thrive, we must embrace the changes and seize opportunities.

Just as Sun Tzu's *Art of War* has been used to prepare a set of competitive guidelines for businesses to follow, we can extend them to fit the evolution of IP telephony and other disruptive and emerging technologies.

Show the way by thinking out of the box and embracing new uses of technology. IP telephony provides a tool that integrates voice and data at a level never before available. It's bringing a new set of tools, so use the new tools to do new things.

Do it right by not embracing technology for the sake of technology. Buy the steak, not the sizzle. Perform business analysis with due diligence, but efficiency. Know the facts. Don't evaluate and plan forever. Know your business. Understand your needs, Evaluate solutions that take your business forward. IP networking is the clear path for business and communications.

IP telephony is a tool, nothing more. But it's a tool that provides a path for equipment vendors, telecommunications service providers, and businesses to innovate and develop integrated solutions that improve efficiency, profitability, and even quality of life for many. Now is the time to use the tools available to our best advantage.

It's critical that we not overlook the obvious. IP telephony has emerged as a solid, viable business tool. Today, it's a disruptive technology that is on the brink of reshaping the entire telecommunications industry. To overlook IP telephony would be a serious mistake, because there will come a day when IP telephony becomes no longer the disruptive upstart, but the sustaining technology. We must continue to look to the future. The one constant trend of the future is change. Technology will continue to evolve: the key is to take advantage of IP telephony at the right time, and that time is now.

APPENDIX A

GLOSSARY

ACELP	Algebraic Code-Excited Linear Prediction
ACK	Acknowledgement
ADPCM	Adaptive Differential Pulse Code Modulation
AMPLIFICATION	Boosting of signal strength
ANALOG	Variable signals across an infinite set of values
ANSI	American National Standards Institute
API	Application Programming Interface
ARP	Address Resolution Protocol
ARPA	Advanced Research Projects Agency
ATM	Asynchronous Transfer Mode
ATTENUATION	Loss in signal strength
BANDWIDTH	The width of the passband used to transmit a signal
BASEBAND	Signals pulsed directly without frequency division
BAUD	Signaling rate
BROADBAND	Signals pulsed using some form of frequency division multiplexing

BROADCAST	Transmit to all members of a group
CHECKSUM	Sum a group of digits for error checking
CIR	Committed Information Rate
CO	Central Office
CODEC	Coder/decoder
COMPANDING	Compressing/expanding
COS	Class of Service
CS-ACELP	Conjugated Structure Algebraic Code-Excited Linear Prediction
CSMA/CD	Carrier Sense Multiple Access with Collision Detection
DDP	Datagram Delivery Protocol
DHCP	Dynamic Host Configuration Protocol
DIFFSERV	Differentiated Services
DIGITAL	Discrete signals acres a predefined set of values
DNS	Domain Name System
DOS	Denial of Service
DSL	Digital Subscriber Line
DSP	Digital Signal Processor
DTMF	Dual Tone Multifrequency
EIA	Electronic Industries Alliance
EP	End point
FCC	Federal Communications Commission
FIFO	First in, first out
FTP	File Transfer Protocol
FULL DUPEX	Two-way
HALF-DUPLEX	Alternating two-way
HOSTID	Host Identifier

HTTP	Hypertext Transfer Protocol
ICMP	Internet Control Message Protocol
IEC	InterExchange Carrier
IEEE	Institute of Electrical and Electronics Engineers
IETF	Internet Engineering Task Force
IHL	Internet Header Length
ILEC	Incumbent Local Exchange Carrier
IP	Internet Protocol
IPDC	IP Device Control
IPX	Internetwork Packet Exchange
ISDN	Integrated Services Digital Network
ISOC	Internet Society
ISP	Internet Service Provider
ISUP	Integrated Services Digital Network User Part
ITSP	Internet Telephony Service Provider
ITU-T	International Telecommunication Union-Telecommunication Standardization Sector
ITXC	Internet Telephony Exchange Carrier
IXC	InterExchange Carrier
LAN	Local Area Network
LD-CELP	Low Delay Code-Excited Linear Prediction
LEC	Local Exchange Carrier
LSR	Label Switching Router
MEGACO	Media Gateway Control
MF	Multifrequency
MGC	Media Gateway Controller
MGCP	Media Gateway Control Protocol

MOS	Mean Opinion Score
MPEG	Moving Pictures Experts Group
MPLS	Multiprotocol Label Switching
MPMLQ	Multipulse Maximum Likelihood Quantization
NAT	Network Address Translation
NETID	Network Identifier
OCTET	Eight bits of data
OS	Operating System
OSI	Open Systems Interconnection
OSPF	Open Shortest Path First
PADDING	Fill characters
PAM	Pulse Amplitude Modulation
PASSBAND	The range of frequencies used to transmit a signal
PBX	Private Branch Exchange
PCM	Pulse Code Modulation
PING	Packet Internet Groper written by Mike Muuss
POP	Point of Presence
PORT	A logical point of connection
POTS	Plain Old Telephone Service
PPP	Point-to-point protocol
PSTN	Public Switched Telephone Network
QOS	Quality of Service
QUANTIZE	Encode a PAM signal into a PCM signal
RAS	Registration, Admission, and Status
RFC	Request For Comments
RIP	Routing Information Protocol

ROUTING	Selecting an appropriate path for a message
RSVP	Resource Reservation Protocol
RTCP	Real Time Control Protocol
RTP	Real Time Transport Protocol
SAP	Session Announcement Protocol
SDH	Synchronous Digital Hierarchy
SDP	Session Description Protocol
SGCP	Simple Gateway Control Protocol
SIMPLEX	One way
SIP	Session Initiation Protocol
SLA	Service Level Agreement
SLIP	Serial Line Internet Protocol
SMTP	Simple Mail Transfer Protocol
S/N	Signal-to-Noise ratio
SNA	Systems Network Architecture
SNMP	Simple Network Management Protocol
SS7	Signaling System 7
SSH	Secure Shell
SWITCHING	Placing temporary links between connections
SYN	Synchronize Sequence Number
TCP	Transmission Control Protocol
TCP/IP	Transmission Control Protocol/Internet Protocol
TDM	Time Division Multiplexing
TIA	Telecommunications Industries Association
TSAPI	Transport Service Access Point Identifier
UDP	User Datagram Protocol
URL	Uniform Resource Locater

VoIP	Voice Over IP
VPN	Virtual Private Network
WAN	Wide Area Network
WIFI	Wireless Fidelity
WWW	World Wide Web
X.400	International standard for email exchange
XML	Extensible Markup Language

APPENDIX B

RESOURCE LIST

When developing a resources list in such a dynamic and fluid technology, inevitably there will be omissions of worthy sites. There will also doubtless be sites included that are invalid by the time readers investigate them. This is not intended to be an all inclusive list, but rather a list of known, viable resources which have shown some resilience in the face of adversity in trying economic times. Some of the listings here are also relatively new, but represent companies that bring a shining light of innovation and growth to the IP telephony market.

When researching the resources available, search engines like Google, and directories like Yahoo! provide readers with a fabulous resource to the latest information.

Solution Vendors

Adir Technologies (www.adirtech.com)—Variety of infrastructure and service solutions for VoIP

Allied Telesyn, Inc. (www.alliedtelesyn.com)—IP video and data services over ADSL

AudioCodes (www.audiocodes.com)—Media gateway solutions for packet networks

Avaya (www.avaya.com)—Complete variety of voice and data networking solutions

Cisco Systems (www.cisco.com)—Comprehensive variety of total networking and VoIP solutions

Clarent (www.clarent.com)—Gateway and softswitch solutions

Comverse (www.comverse.com)—Multimedia messaging, unfied messaging, instant messaging (SMS) solutions

CosmoCom, Inc. (www.cosmocom.com)—IP call center outsourcing and solutions

Digiquant (www.digiquant.com)—Wide selection of wireless, access, and VoIP solutions

EFax.com (www.efax.com)

Estech Systems, Inc. (www.esi-estech.com)—IP phones, PBXs, gateways, and solutions

Genesys Telecom Labs (www.genesyslab.com)—IP Contact Center, an out-of-the-box IP-based platform for mid-market contact centers

Hotsip (www.hotsip.com)—SIP application servers and messaging gateways

IP Unity (www.ipunity.com)—Media and application server technology

JFAX.com (www.jfax.com)—Online fax over IP solutions

Lucent Technologies (www.lucent.com)

Net2Phone (www.net2phone.com)—Business and personal IP telephony solutions

Net6 (www.net6.com)—Network appliance for application integration

Nortel Networks (www.nortel.com)—Comprehensive variety of total networking and VoIP solutions

Onebox.com (www.onebox.com)—Unified message, fax and voice mail

Shoreline Communications (www.goshoreline.com)

Sonexis (www.sonexis.com)—IP PBX and Conference Manager

Voyant Technologies (www.voyanttech.com)—IP telephony technology solutions

VocalTec (www.vocaltec.com)—Softswitch and IP telephony solutions

Voxeo (www.voxeo.com)—XML voice powered solutions

Westwave Communications (www.westwave.com)—Network edge services solutions

Standards Organizations

American National Standards Institute (www.ansi.org)

European Telecommunications Standards Institute (ETSI) (www.etsi.org)

Institute of Electrical and Electronic Engineers (www.ieee.org)

International Softswitch Consortium (www.softswitch.org)

International Telecommunications Union—Telecom Standardization (ITU-T) (www.itu.int/ITU-T)

Internet Engineering Task Force (www.ietf.org)

Telcordia Technologies (formerly BellCore) (www.telcordia.com)

Newsletters and Information Sites

The author publishes a free monthly newsletter that frequently discusses innovations, new companies, and industry trends in a variety of technology segments. Subscriptions and information is available at www.ipadventures.com.

Jeff Pulver publishes a monthly newsletter entitled The Pulver Report. He hosts and regular Voice on the Net (VON events) in different parts of the world. VON conferences have become one of the premier events for IP telephony service and equipment providers. More information on the newsletter and the VON events is available at www.pulver.com.

Steve Taylor and Larry Hettick publish a convergence newsletter that encompasses IP telephony through Network World on a regular basis. www.nwwsubscribe.com/news/scripts/notprinteditnews.asp has a complete listing of all the Network World newsletters available via email.

TMCnet publishes a collection of email newletters, including Internet Telephony eNews. Free subscriptions can be entered at www.tmcnet.com. Users will have to create an account and provide minimal information.

CMP Media publishes Information Week Daily via email at www.informationweek.com. This newsletter covers many areas of technology interest.

General Interest URLs

TelecomWorm.com (www.telecomworm.com) provides an extensive list of telecommunications resources and seems to be kept fairly current.

Gary Kessler, a professor at Champlain College and colleague has published an extensive set of links to voice and data networking resources at www.garykessler.net/library/gck_site.html.

The International Engineering Consortium (IEC) has provided a comprehensive set of tutorials in their WebProForum at www.iec.org/online/tutorials.

The Massachussetts Institute of Technology maintains a program on Internet and Telecom Convergence MIT's Center for Technology, Policy and Industrial Development that discusses the future of the industry, trends and areas of special interest at itel.mit.edu.

Voice over IP Calculator is an interactive resource for engineers and analysts involved in deploying VoIP. They provide a variety of calculators and resources at www.voip-calculator.com.

INDEX

802.11, 228
802.3, 30
802.5, 30
911, 183
μ255, 60
4-Khz voice channel, 122
8, 000 samples per second, 61
802.11 wireless, 209
802.11b, 30, 162, 173–175

access charges, 62, 103
Access Point, 175
Access to new Market Share, 108
ACELP, 65
ACK, 47–49
Acknowledgement, 47
acknowledgement number, 47
Adaptive differential pulse code
    modulation, 65
address resolution protocol, 29
addressing scheme, 26
administrative costs, 99
admission, 157
admission control, 135–146
ADPCM, 65
ADSL, 167, 203
advanced intelligent network, 8, 67, 157
Advanced Research Projects Agency, 28
AIN, 8, 67, 157–158
Alexander Graham Bell, 3
alternate distribution companies, 68
alternate local loop technology, 170

AM, 14
America Online, 172
American National Standards Institute, 56
amplifiers, 4
amplitude, 10
amplitude modulation, 14
analog, 9
analog signals, 9
Analog to Digital Conversion, 57
ANSI, 56
AOL, 105, 170
AOL Instant Messenger, 105
API,
Apple, 77
AppleTalk, 27, 114
Application Layer, 28
application services, 30, 42
ARP, 29
ARPA, 28
ARPANET, 28
ARS, 182
Art of War, 231
AS-400, 114
asymmetric digital subscriber line, 167
asymmetric DSL, 203
AT&T, 3, 54, 66, 77, 156, 170, 213
ATM, 26, 30, 81, 114, 130, 140, 156, 223
attenuation, 11, 25
automatic route selection, 105, 182
Availability, 124, 128–130
average duration, 164
average telephone call, 63

bandwidth, 12–14, 131, 135
barrier to entry, 103
baseband, 25
baud, 14–15
Bearer Data Transport Plane, 87
behavior aggregate, 140
Bell, 3
Bell Labs, 14, 66
Bell South, 213
Bell System, 149, 201
Bellcore, 155
Best effort, 135
best efforts, 31, 41, 49, 123, 131
best path, 133
bicycle principle, 112
billing mechanisms, 157
Binary, 34
bit rate , 14–15
blocking switches, 18
BOOTP, 39
Bootstrap Protocol, 39
broadband, 25
broadcast, 35
buffer, 19
Buffered transmission, 46
building demand, 176
bursty in nature, 123
Business automation tools, 126
Business Evaluations, 106

CA, 87
cable modem, 39
cable modems, 116, 169, 176
cable TV network, 170
call agent, 92–93
call center, 216
call control, 86
call forwarding, 188
Call setup, 77, 83, 92
call waiting, 188
caller ID, 159, 182, 188
Campus calling, 110
Capabilities exchange, 79
capacity utilization, 193

carrier class, 210
CCITT, 56, 60, 72
CCS, 126
cell, 32
central office, 5
central office switch, 7
Centrex, 189, 193
Centrex services, 190
Centrex users, 110
checksum,
Checksum field, 48
CIR,
circuit switched, 62
circuit switched packed data network, 18
circuit switching, 17–19
circuit-switched architecture, 188
Cisco, 68, 77, 86, 140
Class 5 switch, 155, 190, 193–195, 197
Class 5 telco CO switch, 155
Class A, 35–36
Class B, 35–36
Class C, 36
class of service, 124–125
classes, 35
classes of service, 138
classes of traffic, 134
classless addressing, 36
Claude Shannon, 16–17
Clayton Christensen, 213
CLEC, 8, 156, 195
CO, 5, 7
CO infrastructure, 148
CO switch, 62–63, 92–93, 101
codec, 58
command/control, 147
companding, 59–60
competitive local exchange carrier, 8, 195
Computer Telephony Integration, 184, 199
Conference calling, 182
congestion control, 26
Congress, 203
Conjugated structure algebraic-code excited
    linear predictive, 66
connectionless, 20–21, 49, 123

connection-oriented, 20–21, 123
Connection-oriented communication, 46
Connections, 89
connectivity, 176
consulting services, 116
Consumer DSL, 167, 203
Control Flag, 47
Controlled load, 135
converged network infrastructure, 208
convergence, 29, 73, 199, 229
CoS, 138–139, 142
Cost, 124, 128–130
Cost of Doing Business, 99
Cost Reduction, 106
CRC, 26
Create connection, 90
crossbar, 18
Cross-functional teams, 111
CS-ACELP, 65–66, 88
CSMA/CD, 26
CTI, 184
CTO, 199
customer gateway, 194
Customer Satisfaction Improvement, 108
cycles per second, 10
cyclic redundancy check, 26

Darwin's theory of evolution, 231
data, 32
data link layer, 25–26, 42, 46
data network, 118
data warehousing, 230
DDoS, 224
DDP, 27
de facto standard, 30
DEC VAX, 114
delay, 122, 131, 135
delay sensitive, 123
Delete connection, 90
Denial of Service Attacks, 224
dense wave division multiplexing, 162
dense wave division multiplexing, 209
Department of Defense, 28
derived channels, 166–167

derived voice, 205
Design/Operational Envelope, 128–130
Destination Address, 32–33
Destination Port, 46
DHCP, 39–40, 44, 154
dial tone, 101
dialed digits, 189
DID, 153
Differentiated Services, 134, 138–139
Differentiated Services Codepoint, 138
DiffServ, 134, 138, 142
Digit maps, 89
digital, 9
digital loop carrier, 195
digital loop carriers, 166
digital signal, 57
digital signals, 11
Digital subscriber line, 165
digital subscriber line access multiplexer,
    167
Digitization of the PSTN, 167
digitized voice, 103
direct dial long distance, 54
direct inward dialing, 153
discrete multitone, 203
disruptive technologies, 176, 214
disruptive technology, 174, 213–214
distributed call center, 218–219
distributed denial of service, 224
DLC, 166, 195
DMT, 203–204
DNS, 28, 30, 40, 44, 50
Do Not Fragment, 33
domain name system, 28
Domain Names Service, 40, 50
DoS, 224
DoS attacks, 224
dotted decimal, 34
DSCP, 138
DSL, 39, 116, 143, 156, 165, 176
DSL services, 156
DSLAM, 167
DSP,
DSPDN, 18

DTMF, 101
DWDM, 162, 209
dynamic addressing, 38–39
Dynamic Host Configuration Protocol, 39, 154
Dynamic ports, 44

E.164 standards, 81
E911 services, 159, 183
E-commerce, 126
effect of noise on signal, 17
EIA, 56
Email systems, 128
encoding delay time, 64
End point configuration, 90
End points, 89
endpoint, 45
Enterprise Model, 152
enterprise networks, 29
EP, 89
ephemeral ports, 44
equipment costs, 99
Equipment manufacturers, 226
Equipment vendors, 116, 225
Erlang B traffic tables, 126
Erlang-B, 151
ERP systems, 114
error free transmission, 25
Error rate, 124, 128–130
Ethernet, 26, 20, 56, 123, 130, 143, 169
Ethernet phone, 197
ETSI, 155
European Telecommunications Standards Institute, 155
extranet, 29

FAX over IP, 220
FCC, 104
Federal Communications Commission, 104
fiber to the neighborhood, 209
FIFO, 20, 127
File Transfer Protocol, 28
Filtering unwanted frequencies, 61
FIN, 47

Finish, 47
firmphone, 197
first in first out, 20, 127
fixed wireless, 173
Flat rate billing, 164
flat-rate monthly billing, 103
flow specification, 136
FlowSpec, 136
FM, 14
FM radio, 65
foreign exchange office, 194
foreign exchange station, 194
Fourier, 11
four-wire circuits, 8
Fragment Offset, 33
fragmentation, 33
frame, 32
Frame Relay, 30, 81, 114, 130, 140, 156
frames, 25
free space optics, 173
frequency division switching, 18
frequency modulation, 14
FTP, 28, 39, 43, 44
FTTN, 209
full duplex, 8, 25
Full-duple communication, 46
Functional Plane, 87
FXO, 194
FXS, 194

G.711, 60, 65
G.722, 65
G.723.1, 65
G.726, 65
G.728, 65
G.729, 65
g.LITE, 203
gatekeeper, 755–76, 79
gateway, 63, 75, 156
Gateways, 146
Generic Requirement, 303, 155
gigabandwidth, 131, 143
Gigabit Ethernet, 143
Global Crossing, 213

GR-303, 155–156, 195
guarantee delivery, 31, 123
Guaranteed service, 135
guarantees both delivery and sequentiality,
    45
GUI, 179
Guttenberg, 3

H.225, 76, 78
H.245, 76, 79
H.248, 86, 147
H.248/Megaco, 223
H.320, 74
H.323, 45, 75, 80, 85–87, 94–95, 147,
    194–195, 210
H.323 gateway, 146
H.323 protocols, 77
H.323 standards, 156
H.323.Multimedia over packet networks, 74
H.332, 85
half-duplex, 25
Harry Nyquist, 14
header, 32
Header Checksum, 32–33, 46
Hertz (Hz), 10
HFC, 170
Hill Associates, 54
hop-by-hop routing, 141
Hosted IP PBX, 200
HOSTID, 35–36, 38
HP-9000, 114
HR-1542, 176, 208
HTTP, 44, 80–81, 85–86, 125
hub and spoke, 19
hybrid fiber-coax, 170
HyperText Transfer Protocol, 125

IAD, 194, 196–197, 223
IANA, 37
IBM, 124
IBM SNA, 124
ICANN, 38
ICMP, 30, 33
ICQ, 105

IEC, 62, 68, 103–104
IEC switch, 63
IEEE, 55, 173
IETF, 37, 40, 56, 72–73, 80, 133, 138–139
IHL, 32–33
IKE, 45
ILEC, 68, 149, 190
ILEC Gateway, 148
IM, 215
Implement, 117
implementation, 117
Increased Efficiency, 107
Increased Revenue, 107
Incremental improvements, 225
Innovation, 188, 225
Instant messaging , 215
Institute of Electrical and Electronic
    Engineers, 55
integrated access device, 194, 196–197, 223
Integrated Services, 133–134
integrated services digital network, 7
Intel, 77
intellectual capital, 229
interactive voice response, 184
interactive voice response unit, 93
interexchange carrier, 62, 68, 147
International long distance, 110
International Telecommunications Union -
    Telephony sector, 72
Internet, 73–74, 102–105, 123, 125, 126,
    143, 172, 211
Internet Assigned Number Authority, 37
Internet Call Centers, 216
Internet Control Message Protocol, 30
Internet Corporation for Assigned Names
    and Numbers, 38
Internet dot-com bubble, 212
Internet Engineering Task Force, 37, 56,
    72, 133, 138
Internet Header Length, 33
Internet Layer, 42
Internet Phone, 74
Internet Protocol, 1, 17, 26
Internet QoS, 211

Internet service providers, 109
Internet service providers, 38
Internet Society, 56
Internet telephony service provider, 147
Internetwork Packet Exchange, 26
intranet, 29
IntServ, 133–135, 137, 142
IP, 1, 17, 26, 31, 33, 41, 114, 156
IP Address, 90, 102
IP Addresses, 34
IP addressing scheme, 113
IP Centrex, 193, 195–200, 220, 230
IP Device Control, 86
IP Flows, 140
IP network, 42, 146, 154, 157, 193
IP packet, 32, 41, 46, 132
IP packets, 66
IP PBX, 183, 227
IP telephone calls, 101
IP Telephone Service Providers, 103
IP telephony, 50, 62, 67, 101, 104–105, 108,
    113, 115, 117, 143, 146, 152, 178–179,
    210, 215, 228, 231
IP Telephony Gateway, 63, 102, 146,
    153–154
IP telephony system, 112
IP traffic, 123
IP-based PBX, 183
IPDC, 86
IPV4, 41
IPv4 packet, 32
IPv6, 32, 138
IPX, 26–27, 81, 114
ISDN, 7, 114, 178, 190
ISDN videoconferencing standards, 74
ISOC, 56
isochronous Ethernet, 180
IsoEthernet, 180
ISP, 38, 41, 103
ISP LEC, 104
IT staff, 111, 113, 152
ITSP, 103, 147, 150
ITSP Gateway, 149
ITSP model, 149, 151

ITSPs, 103
ITU-T, 56, 72074, 81
ITXC, 152
IVR, 93, 184
IXC, 62

JD Edwards, 114
Jeanne-Baptiste Fourier, 11
Jitter, 131, 135

knowledge management solutions, 230
knowledge work, 163, 178
knowledge workers, 178

Label Distribution Protocol, 141
label switched path, 141
LAN, 26, 42, 83, 140, 147, 173
LAN phone, 197
LAN topologies, 152
LAN-based PBX, 180
law of large numbers, 126
LC-CELP, 65
LDAP, 44
LDP, 141
least cost routing, 105, 182
LEC, 62, 103, 149
legacy technologies, 208
Level3, 150
lifeline service, 203
line powered device, 203
Linux, 180, 210
LMDS, 156, 173
Load coils, 166
local exchange carrier, 62
local exchange switch, 6–7
local loop, 5–6
local loops, 57
local multipoint distribution system, 156,
    173
Local switch, 63
long duration, 122
loss, 11
loss of amplitude, 25
Lotus Notes, 128, 179, 211

Low-delay code excited linear predicate, 65
LSR,
Lucent, 68, 158

MAC address, 42
MAC scheme, 30
Manageability, 124, 128–130
managed service offerings, 189, 227
MCI, 77
mean opinion score, 64
media access control, 30
Media Access Control scheme, 42
Media Gateway, 88
Media Gateway Control Protocol, 87
media gateway controller, 87
Megaco, 86, 89, 90, 93–94, 147
Megaco Event Packages, 91–92
Megaco/H.248, 94–95
MGC, 87–88
Microsoft, 77
Microsoft Exchange, 128
Microsoft NetMeeting, 179
Microsoft Outlook, 179, 211
mission-critical application, 143
mission-critical traffic, 144
MMDS, 156, 174
MMUSIC, 80, 82
Modify connection, 90
monitor traffic, 202
More Fragments, 33
MOS, 64
MPEG,
MPLS, 134, 137, 139–142, 211, 223
MPMLQ, 65–66
MSN Messenger, 105, 179
Mu-law, 60
multichannel multipoint distribution system, 156, 173
Multilocation Centrex, 199
multilocation networks, 109
Multimedia Session Control, 80
Multiprotocol Label Switching, 134, 137, 139, 211

Multipulse maximum likelihood quantization, 66
MultiSystems Interconnect Inc., 128–130

NAT , 37, 40–41
NETID, 35–36, 38
NetMeeting, 45, 105
network address translation, 37
Network addresses, 35
network cost, 99
network gateway, 194
Network Interface Layer, 30, 42
Network Layer, 26–27, 46
network layer protocol, 26–27, 32
Network Performance Envelope, 127
NNTP, 44
no guarantee, 41
nodal delay, 131
nodes, 19
noise, 12, 15–17
Nokia, 140
Nonreal–time, 94
Nortel, 68, 158
Notification request, 90
Nyquist, 14, 122
Nyquist's theorem, 14, 59

OAM&P, 198
OCTET,
octets, 33–34
off hook, 189
offered load, 127
Open logical channel, 79
Open Shortest Path First, 30, 133
open standard, 31
operating system, 28
operational costs, 99
operations, administration, maintenance and provisioning, 198
optical networking, 143
Oracle, 114
OS,
OSI,
OSI Reference Model, 24, 41

OSPF, 30, 33, 133
out of phase, 10

packet, 32
packet network, 103
packet switching, 17
Packetized Voice, 67
packet-over-SONET, 140
packets per second, 151
Padding, 32, 34
pair gain systems, 166
Palm Pilot, 211
PAM, 58
PAM signal, 59
parallel transmission, 25
passband, 12–13
Path Control, 26
PATH message, 136
Payload, 32
PBX, 68, 105, 110, 153–154, 182, 189
PBX extender, 218
PCM, 58, 60, 64–66
PCM voice, 93
PCM-encoded, 88
PC-to-PC telephony, 151
penetration rate for high-speed access,
    177
Per hop behaviors, 140
phase, 10
PHB, 140
physical layer, 25
physically decomposed multimedia
    gateway, 87
pin drop, 65
ping, 30, 43, 125, 172
plain old telephone service, 165
point of presence, 8
policy control, 135–136
POP, 8
POP3, 44
port, 45
port number, 43
port numbers, 43
postal system, 36

POTS, 7, 165, 183
PPP, 30
Presentation Layer, 28
Pringles, 175
prioritization capability, 132
prioritization scheme, 144
Privacy, 201
private addresses, 40
private ports, 44
project manager, 114, 118
Provide New Service, 108
Provisioned quality of service, 134
proxy, 40
proxy server, 81
PSH, 47
PSPDN, 123
PSTN, 3, 7, 14, 28, 47, 57–58, 60, 62, 67,
    72–74, 87, 90, 92, 98, 101, 103, 105,
    122–123, 127, 143, 146, 153, 159, 183,
    190, 211
PSTN providers, 152
PSTN traffic, 102
public IP address, 37
public networks, 40
public switched packet data network, 123
Public Switched Telephone Network, 3
Pulse Amplitude Modulation, 58
pulse code modulation, 58.65
pulse train, 58
Push Function, 47

Q.931, 76, 196
QoS, 31, 33, 124–125, 127, 133
QoS approaches, 142
QoS in IP, 133
QoS issues, 144
QoS mechanisms, 143
quality of service, 31, 33, 124, 151
quantization, 59–60
quantization error, 59
quantizing noise, 59
Quantizing using companding, 61
queuing delay, 18
Qwest, 150, 213

random branching bus , 169
RAS, 76
Real Time Transport Protocol, 75–77, 135
Real-time, 94
real-time traffic, 122
redirect server, 81
regeneration, 12
Registered ports, 44
registrar, 81
registration, 156
registration, admission and status, 76
Reliability, 124, 128–130, 198
reliable transport, 76
Remote terminals, 166
repeaters, 4, 12
Replace the PBX, 109
Replacing the PBX, 179
Request for Comments, 72
Reseaux IP European, 38
Reset connection, 48
Resource Reservation Protocol, 134–135
Resource Reservation Protocol for Traffic Engineering, 137
Response Time, 125, 128–130
return on investment, 106
return on investment, 129
RFC, 72, 134, 138, 140
RIP, 30, 133
RIPE, 38
Robert Heinlein, 109
ROI, 106, 129
routers, 20
routing delay, 131
Routing Information Protocol, 30, 133
routing lookup, 92
routing protocols, 26
RST, 48
RSVP, 134–137, 142
RSVP daemon, 136
RSVP policy control, 136
RSVP-TE, 137
RTCP, 135
RTP, 45, 75–78, 135, 223

S/N, 16
sales engineers, 116
salesperson, 117
Sample the analog signal, 61
SAP, 89
Satellite Internet, 171
SB-ADPCM, 65
SBC, 213
Scalability, 124, 128–130, 198
scope of H.323, 75
SDH, 58
SDP, 82, 86, 89–90
Securities Exchange Commission, 212
Security, 125, 128–130
segment, 46
Senate, 176, 212
Sequence number, 47
serial transmission, 25
service convergence, 131
service delivery parameters, 133
service level agreement, 151
service POP, 209
service provider models, 147
service providers, 116
service set identifier, 175
services, 176
Session Announcement Protocol, 89
session announcements, 82
Session Description Protocol, 82, 89
Session Initiated Protocol, 80, 89
Session Layer, 27
SGCP, 86
Shannon, 16–17, 122
Shannon's Theorem, 16–17
Shoreline Communications, 68
S-HTTP, 85
Signaled quality of service, 133
Signaling Gateway, 88, 93
Signaling Plane, 87
signaling protocol, 137
signaling rate, 14
Signaling System 7, 8, 62
signal-to-noise ratio, 16
Simple Gateway Control Protocol, 86

Simple Mail Transfer Protocol, 28
Simple Network Management Protocol, 50
simplex, 25
sine wave, 9–10, 12
single metric, 133
SIP, 80–82, 85–87, 89, 94–95, 194–195, 210, 223
SIP server, 82, 84
SLA, 151
slash cut, 112
SLIP, 30
Small Business, 115
Small Office/Home Office, 115
SMTP, 28, 39, 44, 80–81
SNA, 26, 124
SNMP, 30, 45, 50
socket, 45
softphone, 197
softswitch, 147, 156, 158, 193
softswitch architecture, 195
Software upgrades, 100
SOHO, 115–116, 162, 165
SONET, 162
Source Address, 32–33
Source and destination port fields, 47
Source Port, 46
space division multiplexing, 18
Splices, 166
spoofed, 224
Sprint, 173
square wave, 11–12
SS7, 862, 74–76, 93
SS7 packet network, 87
SS7 signaling transfer point, 92
SS7 STP, 93
SSH, 85
standards, 54, 56
startup costs, 103
static addressing, 38
statistical multiplexing, 20
step-by-step, 18
store and forward switching, 19–20
STP, 92
Strategic planning, 100

Sun Tzu, 231
Sustaining technologies, 214
sustaining technology, 227
switched network, 5
switching, 17, 32
SYN, 47–49
Synchronize Sequence Number, 47
Synchronous Digital Hierarchy, 58

T.120, 45, 76, 78
T-1, 54, 58, 66
TA-96, 104, 203
Tag Switching, 140
tandem switch, 9
TAPI, 210
Tauzin-Dingell Bill, 176, 208
T-carrier, 54
TCI, 156
TCO, 33
TCP, 27, 30–31, 41, 45, 76–79, 81, 85, 124
TCP segment, 46–47
TCP/IP, 28–30
TCP/IP packet structure, 132
TCP/IP protocol suite, 26, 28–30, 67, 125
TCP/IP stack, 41, 47, 197
TCP/IP suite, 124
TDM, 127, 156
technical consultant, 113
Technical support, 100
telco CO, 149
Telcordia, 86, 155
telecom industry, 212
Telecom service providers, 109
Telecommunications Act of 1996, 104, 149, 176
telecommunications industry, 188–189, 212, 214, 231
Telecommunications Reform Act, 203
telecommunications service industry, 227
Telecommunications Service Providers, 227
telecommuters, 162
telecommuting, 162–163
Telecommuting programs, 177
telephone system, 118

telephony, 182
telephony application programmer's
    interface, 210
Telnet, 30, 45
template, 135
Terabeam, 173
terminal adapter, 199
text-based protocol, 80
The Innovator's Dilemma, 213
three-way handshake, 48
Throughput, 124, 128–130, 135
TIA, 56
tie lines, 98
Tim Berners-Lee, 125
time division multiplexing, 127, 156
time division switches, 18
Time to Live, 32–33
time-division multiplexing, 58
TMG, 87
Token Ring, 26, 30, 56, 130
toll quality voice, 64
ToS field, 132–133, 142
total cost of ownership, 100
Traditional telephony, 62, 215
traffic bearing, 147
traffic engineering, 122
Training, 100
transducer, 4
translates the address, 40
Transmission Control Protocol, 27, 41, 45
Transport Layer, 27, 30, 42–43, 46
Transport Layer protocol, 42
transport service access point, 45
trap, 50
trunking gateway, 147
trunking media gateway, 87
trunks, 6
TSAPI,
TTL, 33
twisted pair, 6
two-wire circuit, 6
Type of Service, 32–33, 132

UDP, 27, 30, 33, 41, 45, 76–79, 81, 85, 90

UDP segment, 50, 90
umbrella standard, 75
unguided light wave, 173
Unified messaging, 223
Unlicensed spectrum, 174
unreliable, 90
unreliable protocol, 31
URG, 47
Urgent Pointer, 47
User agents, 81
User Datagram Protocol, 27, 41

v.34 modem, 204
V5.2, 155
Verizon, 104, 152, 213
Version, 33
virtual PBX, 189
virtual private networks, 188
VocalTec, 74
VoDSL, 203, 205
voice codec, 197
Voice digitization, 54
Voice Mail, 182
Voice mail, 222
Voice Mail over IP, 222
Voice over DSL, 203
voice over IP, 146
Voice Quality, 64
volume, 10
Voyant Technologies, 68
VPN, 114, 188
VT-100, 179

WAP, 174
warchalking, 175
waveform, 9
weak signal, 12
well-known applications, 43
Well-known ports, 44
WEP, 175
WiFi, 20, 56, 162, 173–175, 228
Window field, 48
Windows, 210
Windows NT, 180

WinStar, 173
wireless, 162
wireless access point, 174
wireless equivalent privacy, 175
Wireless Fidelity, 173–175
wireless Internet, 176
Wireless Internet Service Providers
    Association, 173
wireless PDA, 228

WLAN, 173
Work-at-Home, 177
WorldCom, 173, 213
WWW, 30

X.25, 81
X.400, 28
XML, 31

ABOUT THE AUTHOR

Ken Camp is the president and founder of IP Adventures, Inc., a strategic business and technology consulting firm that specializes in helping information professionals use technology to solve business problems. He is also a Consultant Member of Technical Staff with Hill Associates. He lives in Grand Isle, Vermont.